Android
项目开发教程

卓国锋 赵其国 /主编
孟　瑞 刘盼盼 /编著

清华大学出版社
北京

内 容 简 介

本书基于 Android 4.0 版本编写。书中通过一个电子商务项目全面讲解了 Android 开发的过程、技术及应用,包括用户界面布局、服务端通信、基于位置的服务等,涉及的主要知识点从 Activity、Intent 等扩展到 JSON、正则表达式等相关技能。书中将各种知识点融会贯通,随着项目的深入,将基础知识和应用技能的脉络清晰地展现给读者。

本书通过简洁的语言和详细的步骤,帮助读者迅速掌握开发 Android 应用程序所需的基础知识,适合有一定编程经验的读者阅读。书中附有实例与练习。本书可作为高等学校教材,也可供从事 Android 项目开发的人员参考。

本书封面贴有清华大学出版社防伪标签,无标签者不得销售。
版权所有,侵权必究。侵权举报电话:010-62782989 13701121933

图书在版编目(CIP)数据

Android 项目开发教程/卓国锋,赵其国主编. —北京:清华大学出版社,2018(2019.10重印)
ISBN 978-7-302-50900-4

Ⅰ. ①A… Ⅱ. ①卓… ②赵… Ⅲ. ①移动终端-应用程序-程序设计-教材 Ⅳ. ①TN929.53

中国版本图书馆 CIP 数据核字(2018)第 185374 号

责任编辑:焦 虹 李 晔
封面设计:傅瑞学
责任校对:梁 毅
责任印制:刘海龙

出版发行:清华大学出版社
    网 址:http://www.tup.com.cn,http://www.wqbook.com
    地 址:北京清华大学学研大厦 A 座 邮 编:100084
    社 总 机:010-62770175 邮 购:010-62786544
    投稿与读者服务:010-62776969,c-service@tup.tsinghua.edu.cn
    质量反馈:010-62772015,zhiliang@tup.tsinghua.edu.cn
    课件下载:http://www.tup.com.cn,010-62795954
印 装 者:三河市国英印务有限公司
经 销:全国新华书店
开 本:185mm×260mm 印 张:18 字 数:423 千字
版 次:2018 年 12 月第 1 版 印 次:2019 年 10 月第 2 次印刷
定 价:39.80 元

产品编号:062177-01

前言

随着 Android 系统的迅速发展和普及，整个 Android 系统的生态环境也在逐渐成熟，其应用不仅局限于手机，在平板电脑、智能穿戴、电子商务等方面也在迅速发展。

1. 本书讲解的主要内容

本书重点讲解 Android 系统功能模块涉及的主要知识点，并通过技能扩展讲解常用的实用技能。本书的项目原型是公司电子商务的基础版本，有助于读者了解一个真实项目的整个开发流程。全书共 13 章。其中，第 1~3 章介绍 Android 的历史、现状及书中项目的背景；第 4~11 章介绍项目的主要功能模块和重要知识点；第 12~13 章介绍如何发布 Android 应用，并介绍 HTML 5 的基础知识。

2. 本书适合的读者

首先，要求读者具备 Java 语言的基础知识；其次，要理解书中讲到的知识点。"好记性不如烂笔头"，只有不断练习，才能把学到的知识变成自己的能力。

在学习项目开发的时候，可以边学、边练，这样在学习完成后就可具有一定的项目实践经验，而不是仅了解一些空洞的、概念的东西。具有一定的基础后，可尝试阅读 API 开发文档、项目源码及 Android 的系统源码，这对于 Android 的学习是非常重要的。在学习与实践的过程中，不知不觉，编程就会变成一件快乐的事情。

本书由成都职业技术学院教师编写。赵其国负责编写第 1、2、3、4 章，共 13.1 万字。卓国锋负责编写第 5、6、7、8、9、10 章，共 25.8 万字。孟瑞负责编写第 11、12 章，共 2.7 万字。刘盼盼负责编写第 13 章及附录 A 与附录 B，共 0.7 万字。

编 者

目录

第 1 章 Android 应用开发概述 ... 1

1.1 Android 应用开发的历史与现状 ... 1
1.2 Android 应用基本架构 ... 2
    1.2.1 Android 系统介绍 ... 2
    1.2.2 Android 平台架构及特性 ... 3
1.3 Android 应用开发的特点 ... 4
    1.3.1 Android 应用的组成 ... 4
    1.3.2 Android 堆栈管理 ... 5
    1.3.3 Android 生命周期 ... 6
    1.3.4 Android 布局管理 ... 8
    1.3.5 Activity 交互 ... 8
    1.3.6 SQLite ... 9
    1.3.7 Android 实际开发经验分享 ... 9
1.4 Android 开发工具简介 ... 9
    1.4.1 下载和安装 JDK ... 9
    1.4.2 安装 Eclipse ... 12
    1.4.3 安装 ADT 插件 ... 12
1.5 知识点与技能回顾 ... 16
1.6 练习 ... 16

第 2 章 为开发做好准备 ... 17

2.1 手机客户端准备 ... 17
2.2 网络环境准备 ... 17
2.3 服务器准备 ... 17
    2.3.1 安装并配置 Tomcat ... 17
    2.3.2 安装并配置 MySQL ... 19
    2.3.3 Navicat 的安装和使用 ... 28

2.4 知识点与技能回顾 …………………………………………………………… 30
2.5 练习 …………………………………………………………………………… 30

第3章 MeDemo 项目介绍 ……………………………………………………… 31

3.1 项目背景 ……………………………………………………………………… 31
3.2 项目需求分析 ………………………………………………………………… 31
3.3 项目用例分析 ………………………………………………………………… 31
3.4 项目流程 ……………………………………………………………………… 33
3.5 项目数据库 …………………………………………………………………… 35
3.6 项目时序图 …………………………………………………………………… 35

第4章 用户注册 …………………………………………………………………… 37

4.1 用户注册总体设计 …………………………………………………………… 37
4.2 数据库的准备 ………………………………………………………………… 38
4.3 用户注册重要知识点详解 …………………………………………………… 39
    4.3.1 Android 项目目录结构 ……………………………………………… 39
    4.3.2 xml 布局文件的创建 ………………………………………………… 40
    4.3.3 Activity 的创建 ……………………………………………………… 50
    4.3.4 dimen 资源文件 ……………………………………………………… 55
    4.3.5 drawable 资源文件 …………………………………………………… 56
    4.3.6 客户端与服务器的交互 ……………………………………………… 57
4.4 用户注册 ……………………………………………………………………… 57
    4.4.1 用户注册的具体实现 ………………………………………………… 57
    4.4.2 几个关键的类 ………………………………………………………… 71
    4.4.3 AndroidManifest.xml ………………………………………………… 76
4.5 用户注册功能的调试 ………………………………………………………… 78
4.6 知识点回顾与技能扩展 ……………………………………………………… 79
    4.6.1 知识点回顾 …………………………………………………………… 79
    4.6.2 技能扩展 ……………………………………………………………… 79
4.7 练习 …………………………………………………………………………… 85

第5章 用户登录 …………………………………………………………………… 86

5.1 用户登录总体设计 …………………………………………………………… 86
5.2 用户登录的实现 ……………………………………………………………… 87
    5.2.1 登录的具体实现 ……………………………………………………… 87
    5.2.2 客户端与服务器的交互 ……………………………………………… 93
    5.2.3 后台服务接口文档 …………………………………………………… 94

5.3 用户登录的调试 ··· 94
5.4 支持用户使用第三方账号登录 ····································· 95
 5.4.1 什么是第三方登录 ··· 95
 5.4.2 第三方账号登录方式 ······································ 95
 5.4.3 使用第三方账号登录 ······································ 95
5.5 知识点回顾与技能扩展 ··· 96
 5.5.1 知识点回顾 ··· 96
 5.5.2 技能扩展 ·· 96
5.6 练习 ··· 107

第 6 章 向用户展示内容 108

6.1 基本内容展示总体设计 ·· 108
6.2 数据库准备 ·· 109
 6.2.1 数据库商户 ··· 109
 6.2.2 数据库商户表 ·· 109
 6.2.3 后台服务端接口文档 ····································· 110
6.3 内容展示知识点详解 ·· 112
 6.3.1 Fragment 介绍 ··· 112
 6.3.2 FragmentManage 介绍 ··································· 115
 6.3.3 FragmentTransaction 介绍 ······························ 115
6.4 内容展示 ·· 116
 6.4.1 内容展示的具体实现 ····································· 116
 6.4.2 客户端和服务端交互 ····································· 132
6.5 知识点回顾与技能扩展 ·· 133
 6.5.1 知识点回顾 ··· 133
 6.5.2 技能扩展 ·· 133
6.6 练习 ··· 146

第 7 章 支持用户基于 LBS 的应用 147

7.1 用户定位 ·· 147
 7.1.1 LBS 与常见第三方地图服务简介 ······················ 147
 7.1.2 在地图上找到自己 ·· 148
7.2 摇一摇 ··· 158
 7.2.1 摇一摇功能的实现 ·· 158
 7.2.2 传感器介绍 ··· 161
7.3 知识点回顾 ··· 162
7.4 练习 ··· 163

第 8 章　用户搜索与结果展示 · 164

- 8.1　用户搜索功能总体设计 · 164
- 8.2　用户搜索功能知识点详解 · 165
- 8.3　用户搜索的实现 · 167
- 8.4　知识点回顾 · 177
- 8.5　练习 · 177

第 9 章　与用户互动 · 178

- 9.1　让用户参与评价 · 178
  - 9.1.1　用户发表评价的界面 · 178
  - 9.1.2　用户发表评价 · 179
  - 9.1.3　商户的评价列表展示 · 190
- 9.2　让用户分享 · 191
  - 9.2.1　什么是分享 · 191
  - 9.2.2　让用户将内容分享到社交平台 · 191
- 9.3　给用户推送消息 · 197
  - 9.3.1　推送的几种常见解决方案 · 197
  - 9.3.2　常用的推送平台 · 197
- 9.4　知识点回顾 · 209
- 9.5　练习 · 209

第 10 章　添加商户信息 · 210

- 10.1　添加商户信息总体设计 · 210
- 10.2　商户数据库准备 · 211
- 10.3　Intent 详解 · 211
- 10.4　添加商户信息流程控制 · 214
- 10.5　知识点回顾与技能扩展 · 224
  - 10.5.1　知识点回顾 · 224
  - 10.5.2　技能扩展 · 224

第 11 章　让用户使用体验更佳 · 234

- 11.1　用户手机网络环境 · 234
- 11.2　知识点回顾 · 235

第 12 章　发布和管理 Android 应用 · 236

- 12.1　为何要发布 · 236

12.2 在哪里发布 ………………………………………………………………… 236
12.3 如何发布到第三方市场 ……………………………………………………… 236
    12.3.1 在 Eclipse 中对 Android 应用签名 …………………………………… 236
    12.3.2 发布 APK 到第三方市场 ……………………………………………… 240
12.4 版本与版本管理 …………………………………………………………… 243
    12.4.1 设置版本号和版本名 ………………………………………………… 243
    12.4.2 获取当前版本信息 …………………………………………………… 243
12.5 如何让用户升级 …………………………………………………………… 243
    12.5.1 服务器准备 …………………………………………………………… 243
    12.5.2 客户端实现 …………………………………………………………… 244
12.6 知识点回顾 ………………………………………………………………… 252

第 13 章 与用户终端设备无关的 HTML 5 …………………………………… 253

13.1 什么是 HTML 5 …………………………………………………………… 253
    13.1.1 综述 …………………………………………………………………… 253
    13.1.2 发展历史 ……………………………………………………………… 253
    13.1.3 特性 …………………………………………………………………… 254
    13.1.4 未来趋势 ……………………………………………………………… 255
13.2 用 HTML 5 实现内容展示 ………………………………………………… 257
    13.2.1 WebView 组件 ………………………………………………………… 257
    13.2.2 HTML 5 本地存储 …………………………………………………… 260
    13.2.3 HTML 5 的地理位置服务 …………………………………………… 264

附录 A AndroidManifest.xml 中的权限 ……………………………………… 266

附录 B Intent 和 Action 汇总 ………………………………………………… 269

参考文献 ……………………………………………………………………… 275

第1章 Android 应用开发概述

1.1 Android 应用开发的历史与现状

全球智能手机销量已超过 PC 的销量,"计算设备移动化"的时代即将到来。

Android 一词的本义指"机器人",同时也是 Google 于 2007 年 11 月 5 日宣布的基于 Linux 平台的开源手机操作系统的名称。该平台由操作系统、中间件、用户界面和应用软件组成,号称是首个为移动终端打造的真正开放和完整的移动软件。

随着 Android 系统的迅速壮大,它已经成为全球范围内最具有广泛影响力的操作系统。手机、平板电脑、电视、可佩戴手环等设备的用量在爆发式地增长,同时 Android 也成为这些设备的首选。在这样的背景下,Android 软件人才的需求在不断增长,Android 开发人才正在快速成长。

自 Android 系统首次发布至今,Android 经历了很多次版本更新。表 1.1 列出了 Android 系统的不同版本的发布时间及对应的版本号。

表 1.1 Android 版本发布日期及版本号

Android 版本	发 布 日 期	代　　号
Android 1.1	2008 年 9 月	无
Android 1.5	2009 年 4 月 30 日	Cupcake(纸杯蛋糕)
Android 1.6	2009 年 9 月 15 日	Donut(炸面圈)
Android 2.0/2.1	2009 年 10 月 26 日	Eclair(长松饼)
Android 2.2	2010 年 5 月 20 日	Froyo(冻酸奶)
Android 2.3	2010 年 12 月 6 日	Gingerbread(姜饼)
Android 3.0/3.1/3.2	2011 年 2 月 22 日	Honeycomb(蜂巢)
Android 4.0	2011 年 10 月 19 日	Ice Cream Sandwich(冰激凌三明治)
Android 4.1/4.2/4.3	2012 年 6 月 28 日	Jelly Bean(果冻豆)
Android 4.4	2013 年 9 月 4 日	KitKat(奇巧)
Android 5.0/L	2014 年 6 月 26 日	Lemon Meringue Pie(柠檬酥皮馅饼)

Android 系统迅猛发展的同时，系统碎片化也成为 Android 生态里最为严峻的问题。Google 公司为了解决这些问题，版本不断优化。Google 公司公布的 Android 版本分布如图 1.1 所示。

版本号	版本名	API	市场占有率/%
2.2	Froyo	8	0.8
2.3.3-2.3.7	Gingerbread	10	14.9
4.0.3-4.0.4	Ice Cream Sandwich	15	12.3
4.1.x	Jelly Bean	16	29.0
4.2.x		17	19.1
4.3		18	10.3
4.4	KitKat	19	13.6

数据收集截至时间2014-07-04，数据低于0.1%的版本没有显示

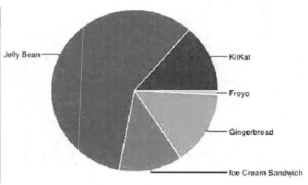

图 1.1 Android 版本分布

从图 1.1 可以看出，从 Android 4.0 开始，Android 系统有了一个质的飞跃，用户绝大部分都已经升级为 4.0 以上，碎片化的问题在很大程度上得到了解决。本书将重点讲解 Android 4.0 及以上版本的开发，功能向下兼容最低到 Android 2.2 版本。

1.2 Android 应用基本架构

1.2.1 Android 系统介绍

Android 系统的底层是建立在 Linux 系统之上的，它采用了软件堆层（Software Stack，又名软件叠层）的架构，主要分为三部分：低层以 Linux 核心工作为基础，只提供基本功能，其他应用软件则由各公司自行开发，以 Java 作为编写程序的一部分。另外，为了推广此技术，Google 和其他几十个手机公司建立了开放手机联盟（Open Handset Alliance）。例如，Hero 的 UI 即由 HTC 自行研发，名为 Senes，之前没有一款 Android 手机有如此华丽、人性化的界面。

与 iPhone 相似，Android 采用 WebKit 浏览器引擎，具备触摸屏、高级图形显示和上网功能。用户能够在手机上查看电子邮件、搜索网址和观看视频节目等，比 iPhone 等其他手机更强调搜索功能，界面更强大，可以说是一种融入全部 Web 应用的单一平台。

作为一个手机平台，Android 在技术上的优势主要有以下几点：
- 全开放智能手机平台。
- 多硬件平台的支持。
- 使用众多标准化技术。
- 核心技术完整、统一。
- 完善的 SDK 和文档。

- 完善的辅助开发工具。

Android 的开发者可以在完备的开发环境中进行开发。Android 的官方网站提供了丰富的文档、资料,这样有利于 Android 系统的开发,使其运行在一个良好的生态环境中。

1.2.2 Android 平台架构及特性

从宏观的角度来看,Android 是一个开放的软件系统,它包含了众多源代码。从下至上,Android 系统分成 4 个层次:
- 第 1 层次:Linux 操作系统及驱动。
- 第 2 层次:本地代码(C/C++)框架。
- 第 3 层次:Java 框架。
- 第 4 层次:Java 应用程序。

Android 系统架构如图 1.2 所示。应用程序的开发主要关注第 3 层次和第 4 层次之间的接口。Android 应用程序由一个或多个组件组成。

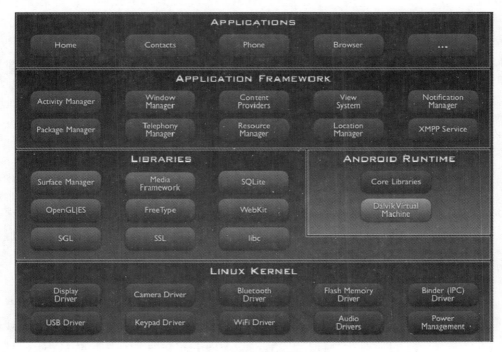

图 1.2　Android 系统架构

1. 活动

具有可视化 UI 的应用程序是用活动(Activity)实现的。当用户从主屏幕或应用程序启动器选择一个应用程序时,就会开始一个动作。

2. 服务

服务(Service)应该用于需要持续较长时间的应用程序,例如网络监视器或更新检查应用程序。

3. 内容提供程序

可以将内容提供程序(ContentProvider)看作数据库服务器。内容提供程序的任务是管理对持久数据的访问,例如 SQLite 数据库。如果应用程序非常简单,那么可能不需要创建内容提供程序。如果要构建一个较大的应用程序,或者构建需要为多个活动或应用程序提供数据的应用程序,那么可以使用内容提供程序实现数据访问。

4. 广播接收器

Android 应用程序可用于处理一个数据元素,或者对一个事件(例如接收文本消息)作出响应。Android 应用程序是连同 AndroidManifest.xml 文件一起部署到设备的。AndroidManifest.xml 包含必要的配置信息,以便将它适当地安装到设备上。它包括必需的类名和应用程序能够处理的事件类型,以及运行应用程序所需的许可。例如,如果应用程序需要访问网络(例如要下载一个文件),那么 AndroidManifest.xml 文件中必须显式地列出该许可。很多应用程序可能启用了这个特定的许可,这样有助于减少恶意应用程序损害设备的可能性。1.3 节讨论构建 Android 应用程序所需的开发环境。

1.3 Android 应用开发的特点

1.3.1 Android 应用的组成

Android 的应用一般由四个关键部分组成：Activity、IntentReceiver、Service、ContentProvider。

1. 组件注册

基本组件都需要注册才能使用,每个 Activity、Service、ContentProvider 都需要在 AndroidManifest.xml 文件中进行配置,AndroidManifest 文件中未声明的 Activity、服务以及内容提供程序将不为系统所见,从而也就不可用,而 BroadcastReceive 广播接收者的注册分静态注册(在 AndroidManifest.xml 文件中进行配置)和通过代码动态创建并以调用 Context.registerReceiver()的方式注册至系统。需要注意的是：在 AndroidManifest 文件中进行配置的广播接收者会随系统的启动而一直处于活跃状态,只要接收到感兴趣的广播就会触发(即使程序未运行)。

在 AndroidManifest.xml 文件中进行注册如下。

<activity>元素的 name 属性指定了实现这个 activity 的 Activity 的子类。icon 和 label 属性指向了包含展示给用户的此 activity 的图标和标签的资源文件。

<service> 元素用于声明服务。

<receiver> 元素用于声明广播接收器。

<provider> 元素用于声明内容提供者。

2. 组件激活

内容提供者的激活：当接收到 ContentResolver 发出的请求后，内容提供者被激活。而其他三种组件——Activity、服务和广播接收器被叫做 Intent 的异步消息所激活。

- Activity 的激活通过传递一个 Intent 对象至 Context.startActivity() 或 Activity.startActivityForResult() 以载入（或指定新工作给）一个 Activity。相应的 Activity 可以通过调用 getIntent() 方法来查看激活它的 Intent。如果期望它所启动的那个 Activity 返回一个结果，则会以调用 startActivityForResult() 来取代 startActivity()。例如，启动另一个 Activity 使用户挑选一张照片后，也许想知道哪张照片被选中了。结果将会被封装在一个 Intent 对象中，并传递给发出调用的 Activity 的 onActivityResult() 方法。
- Service 的激活可以通过传递一个 Intent 对象至 Context.startService() 或 Context.bindService() 实现。前者 Android 调用 Service 的 onStart() 方法将 Intent 对象传递给它；后者 Android 调用 Service 的 onBind() 方法将这个 Intent 对象传递给它。
- 发送广播可以传递一个 Intent 对象至 Context.sendBroadcast()。
- 对于 Context.sendOrderedBroadcast() 或 Context.sendStickyBroadcast()，Android 会调用所有对此广播有兴趣的广播接收器的 onReceive() 方法，将 Intent 传递给它们。

3. 组件关闭

内容提供者仅在响应 ContentResolver 提出请求的时候激活。广播接收器仅在响应广播信息的时候激活，所以没有必要去显式地关闭这些组件。

关闭 Activity：可以通过调用它的 finish() 方法来关闭一个 Activity。

关闭 Service：对于通过 startService() 方法启动的 Service，要调用 Context.stopService() 方法关闭 Service；对于使用 bindService() 方法启动的 Service，要调用 Contex.unbindService() 方法关闭 Service。

1.3.2　Android 堆栈管理

任务其实就是 Activity 堆栈，它由一个或多个 Activity 组件共同完成一个完整的用户体验。Activity 堆栈如图 1.3 所示。

Activity 堆栈的原理：先进后出。

一个具有此功能的 Activity1，调用 startActivity() 方法启动 Activity2。Activity1 在

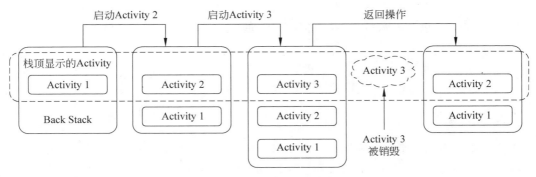

图 1.3 Activity 堆栈

未调用 finish() 的情况下, 被压入栈底, Activity2 则展现在顶部。此时堆栈中有两个 Activity。

同理, 若 Activity2 调用 startActivity() 方法启动 Activity3, 则 Activity2 和 Activity1 被压到栈底, Activity3 在栈顶, 展示在用户面前。

当按下 Backspace 键的时候, 当前 Activity3 被销毁, 移出栈中, Activity2 展现在用户面前。最后 Activity1 也被销毁, 整个应用退出。

1.3.3 Android 生命周期

1. Activity 的生命周期

堆栈管理着 Activity, 每个应用都是必须要有的。Activity 代表一个应用的一个具体的界面管理类, 其本身并不显示。Activity 管理着这个界面里的各种组件, 它本身也有生命周期, 如图 1.4 所示。

Activity1 的启动顺序: onCreate()→onStart()→onResume()。

启动 Activity2 时: Activity1 onPause()→Activity2 onCreate()→Activity2 onStart()→Activity2 onResume()→Activity1 onStop()。

返回 Activity1 时: Activity2 onPause()→Activity1 onRestart()→Activity1 onStart()→Activity1 onResume()→Activity2 onStop()→Activity2 onDestroy()。

2. IntentReceiver 的生命周期

IntentReceiver 可使应用对外部事件作出响应。例如, 当应用正在执行时, 突然来了电话, 这个时候可使用 IntentReceiver 作出处理, 从而使应用更加健壮。

IntentReceiver 的生命周期只有 10 秒。如果在 onReceive() 内做超过 10 秒的事情, 就会显示 ANR(Application No Response) 程序无响应的错误信息。

它的生命周期为从回调 onReceive() 方法开始到该方法返回结果后结束。

3. Service 的生命周期

Service 的生命周期是由 Android 系统来决定的, 而不是由具体应用的线程左右。当

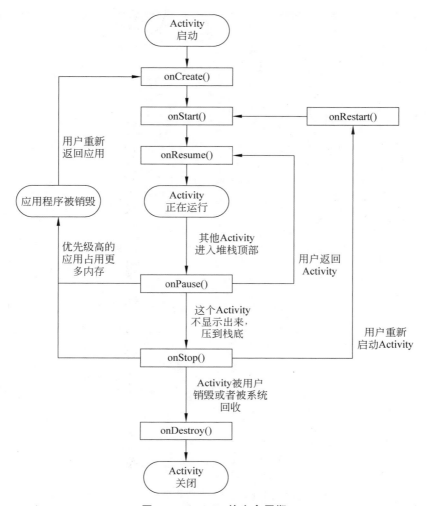

图 1.4 Activity 的生命周期

应用要求在没有界面显示的情况还能正常运行时（要求有后台线程，而后台线程不会被系统回收，直到线程结束），就需要用到 Service。

Service 完整的生命周期是：从调用 onCreate()开始直到调用 onDestroy()结束，如图 1.5 所示。

Service 有两种使用方法如下：

（1）以调用 Context. startService()启动，以调用 Context. stopService()结束。

Service startService()→Service onCreate()→Service onStart()→Service running→Service stopService()→Service onDestroy()→Service stop。

（2）以调用 Context. bindService()方法建立，以调用 Context. unbindService()关闭。

Service bindService()→Service onCreate()→Service onBind()→Service running→Service onUnbind()→Service onDestroy()→Service stop。

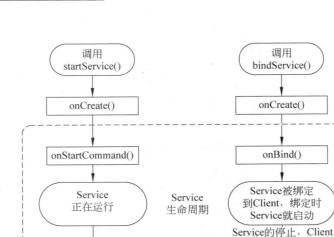

图 1.5　Service 的生命周期

1.3.4　Android 布局管理

FrameLayout：左上角只显示一个组件。

LinearLayout：线性布局管理器，分为水平和垂直两种，只能进行单行布局。

TableLayout：任意行和列的表格布局管理器。其中 TableRow 代表一行，TableRow 的每一个视图组件代表一个单元格。

AbsoluteLayout：绝对布局管理器。坐标轴的方式是：左上角是原点(0,0)，X 轴向右递增，Y 轴向下递增。

RelativeLayout：相对布局管理器。根据最近的一个视图组件，或是顶层父组件来确定下一个组件的位置。

1.3.5　Activity 交互

Intent 的中文为"意图"，主要处理 Activity、Service、Receiver 和 Provider 之间的交互。举个简单的例子，Intent.ACTION_CALL 可调出用户默认的拨打电话页面。

SharedPreferences 是 Android 平台上一个轻量级的存储类，主要保存一些常用的配置。SharedPreferences 类似过去 Windows 系统上的 ini 配置文件，它分为多种权限，可以全局共享访问，最终以 xml 文件的方式来保存。虽然它的效率不如 Intent，但是由于可以共享，所以可以在 Activity 之间交互，其效率比 SQLite 要高。

SQLite 也就是数据库，不推荐这种方式，因为其效率问题。在不同应用之间交互，或在永久存储的情况下，也可以考虑使用 SQLite。

1.3.6 SQLite

SQLite 是 Android 中提供的内置数据库，据说比 MySQL 更轻巧。SQLite 也是开源产品，可以用 SQL 语句直接操作。插入、更新、删除都可以直接用 SQL 语句，调用 execSQL()就可以了，而查询需要使用 rawQuery()来完成。查询结果返回一个可滚动的结果集，在对 Cursor 操作前，需要将其游标移动到第一位，每取一个结果向下移一位。

1.3.7 Android 实际开发经验分享

（1）自定义组件的显示问题。在写自定义 View 的时候经常要对视图的 X、Y 进行调整，以达到预期的理想位置。可以将每个组件的 X、Y 坐标值画到组件旁边，这样很直观，一看就知道该怎样调整。

（2）使用 Log 来打印日志和进行调试。

（3）使用 LogCat 视图。在 showView 中有 LogCat 视图。LogCat 视图会显示一些 Android 仿真器打印出的堆栈信息，对应用的调试非常有帮助。另外 Log 打印的日志也在这里显示。

（4）使用 Emulator Control 视图。Emulator Control 可以完成一些简单的设备操作，例如模拟来电、短消息。

（5）灵活运用 tools 目录中的工具。该目录在 Android SDK 中提供。通过这些工具可以操作 Android 仿真器。例如创建一个虚拟 SD 卡，将系统中的文件移动到虚拟 SD 卡中。Android 提供了相应的工具，相关命令可以上网查阅。

1.4 Android 开发工具简介

1.4.1 下载和安装 JDK

下载和安装 JDK 的步骤如下：

（1）打开 http://www.oracle.com/technetwork/java/javase/downloads/index.html（见图 1.6），单击 JDK 下的 DOWNLOAD 按钮进入下载页面（见图 1.7）。

（2）运行下载的 jdk-7xx-xxx-x64.exe 文件，安装 JDK。

（3）配置 JDK 环境，单击鼠标右键，选择"计算机"→"属性"→"高级系统设置"，配置 JAVA_HOME 变量（见图 1.8）。

（4）配置 classpath。选中"系统变量"，查看是否有 classpath 项目。如果没有，就单击"新建"；如果已经存在，就选中 classpath 选项。单击"编辑"按钮，然后在"变量名"中填写"classpath"，在"变量值"中填写".;%JAVA_HOME%\lib"，一定需要".;"。

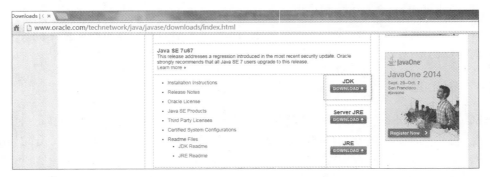

图 1.6　JDK 下载页面

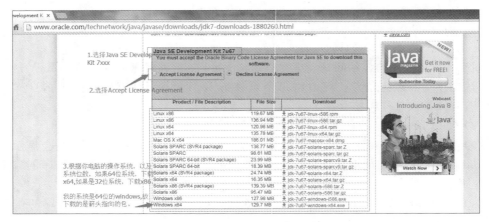

图 1.7　JDK 系统和版本选择

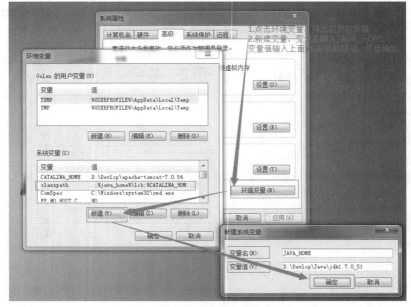

图 1.8　JDK 环境变量配置

（5）现在可以进行 path 的配置了。与设定 classpath 时类似，在"变量名"输入框中填写"path"，在"变量值"输入框中填写"%JAVA_HOME%\bin"。

（6）JDK 的环境变量已经配置完成。打开命令提示符窗口，输入命令"java -version"，通过 Java 版本的信息，来确定安装是否成功。首先单击"开始"，然后单击"运行"，进入 cmd 命令行后，输入"java -version"，出现如图 1.9、图 1.10 所示的结果，表明 JDK 环境配置成功。

图 1.9　调用 cmd 脚本命令窗口

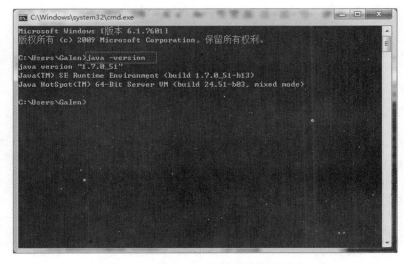

图 1.10　检查 JDK 环境是否配置成功

1.4.2　安装 Eclipse

安装 Eclipse 的步骤如下：

（1）打开官网 eclipse.org，单击菜单栏上面的 download。

（2）若计算机是 32 位，则选择 32 位版本的 Eclipse；若计算机是 64 位，则选择 64 位版本的 Eclipse，如图 1.11 所示。

（3）进入下载页面，一般单击红框里面的网址就可以下载了，如图 1.12 所示。

（4）下载完成后，解压到放置软件的目录中，如图 1.13 所示，单击 eclipse.exe 即可启动。

图 1.11　Eclipse 下载页面（1）

图 1.12　Eclipse 下载页面（2）

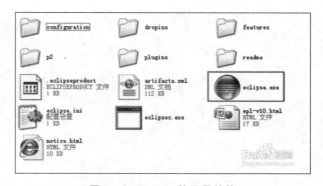

图 1.13　Eclipse 的目录结构

1.4.3　安装 ADT 插件

安装 ADT 插件的步骤如下：

(1) 打开 Eclipse,选择菜单项 Help→Install New Software,如图 1.14 所示。

(2) 安装 ADT 插件的过程如图 1.15、图 1.16、图 1.17、图 1.18、图 1.19、图 1.20、图 1.21、图 1.22 所示。

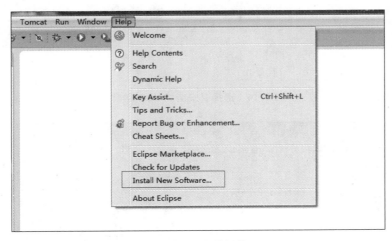

图 1.14 安装插件

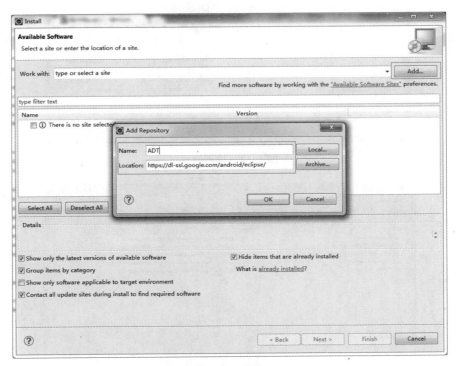

图 1.15 添加 ADT 插件下载站点

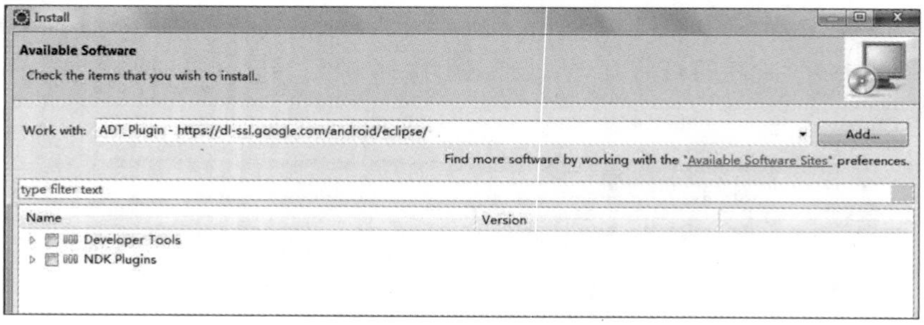

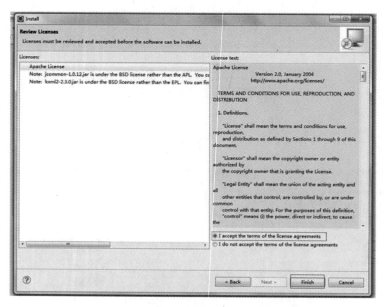

图 1.16　选择要安装的 ADT 插件

图 1.17　使用协议

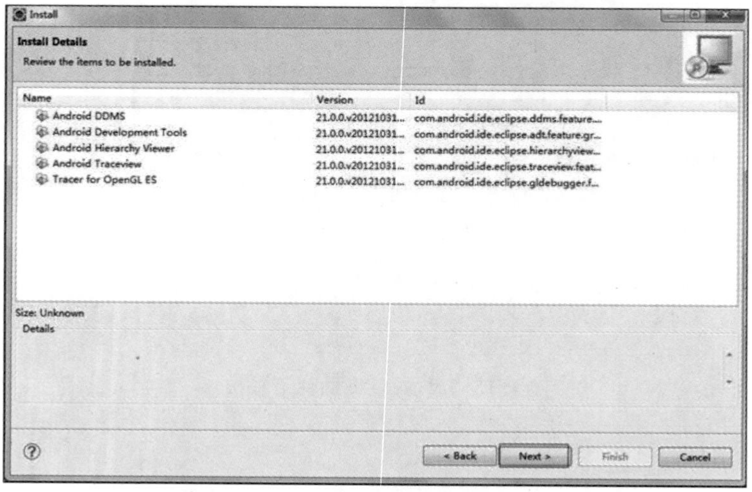

图 1.18　安装进度提示

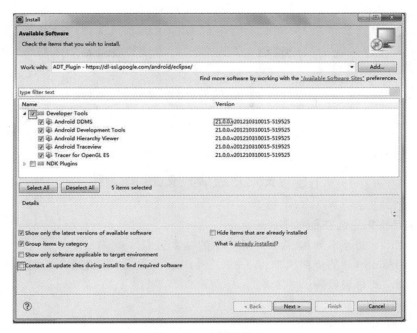

图 1.19 安装的版本

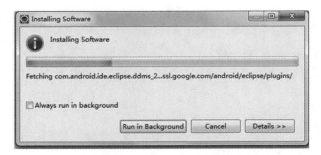

图 1.20 安装进度条

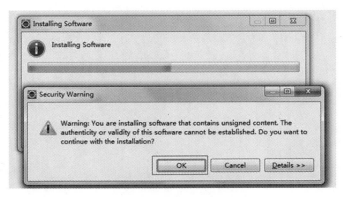

图 1.21 提示是否重启 Eclipse

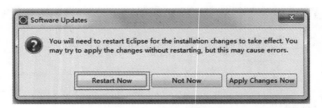

图 1.22　立即重启 Eclipse

1.5　知识点与技能回顾

本章主要知识点与技能如下：
（1）Android 的基本架构、Android 应用的基本组成。
（2）Android 组件的生命周期，Activity、Service、Reciver 以及所涉及的 Fragment。
（3）Android 布局管理。
（4）Android 交互。
（5）Android 堆栈管理。

1.6　练　　习

安装 JDK、Eclipse 及 ADT 插件。

第 2 章 为开发做好准备

2.1 手机客户端准备

手机客户端的准备步骤如下：
(1) 准备一部 Android 手机（Android 系统版本在 2.3 以上）。
(2) 手机开启 USB 调试。根据手机 Android 的系统版本，采取不同的方式开启。
- Android 2.x 版本：选择手机设置→应用程序→开发，勾选 USB 调试。
- Android 4.0.x、4.1.x 版本：选择手机设置→开发人员选项，勾选 USB 调试。
- Android 4.2.x 及以上版本：选择手机设置→关于手机→版本号（连续单击 5～7 次）→返回（设置）→开发者选项，勾选 USB 调试。

(3) 在计算机中安装手机 USB 驱动。
(4) 将手机连接计算机，打开 Eclipse，选择 DDM 查看是否检测到接入设备。

2.2 网络环境准备

手机连接的 WiFi 和服务端连接的网络在同一局域网的条件下，安装 Tomcat 和 MySQL，并进行配置。

2.3 服务器准备

2.3.1 安装并配置 Tomcat

安装并配置 Tomcat 的步骤如下：
(1) 如果没安装 JDK，请参照 1.4.1 节的方法先安装 JDK。安装好 JDK 后，首先到 Tomcat 官网 http://tomcat.apache.org/，根据自己的计算机系统对应下载 32 位或 64 位版本的 Tomcat，如图 2.1 所示。
(2) 下载完成后，解压到一个文件夹中。进入其中的 bin 文件夹，如图 2.2 所示。
(3) 运行 startup.bat 启动 Tomcat，如图 2.3 所示。

（4）启动浏览器，访问 http：//localhost：8080/。若出现如图 2.4 所示的界面，则表明 Tomcat 安装成功。

图 2.1　在官网下载 Tomcat 安装包

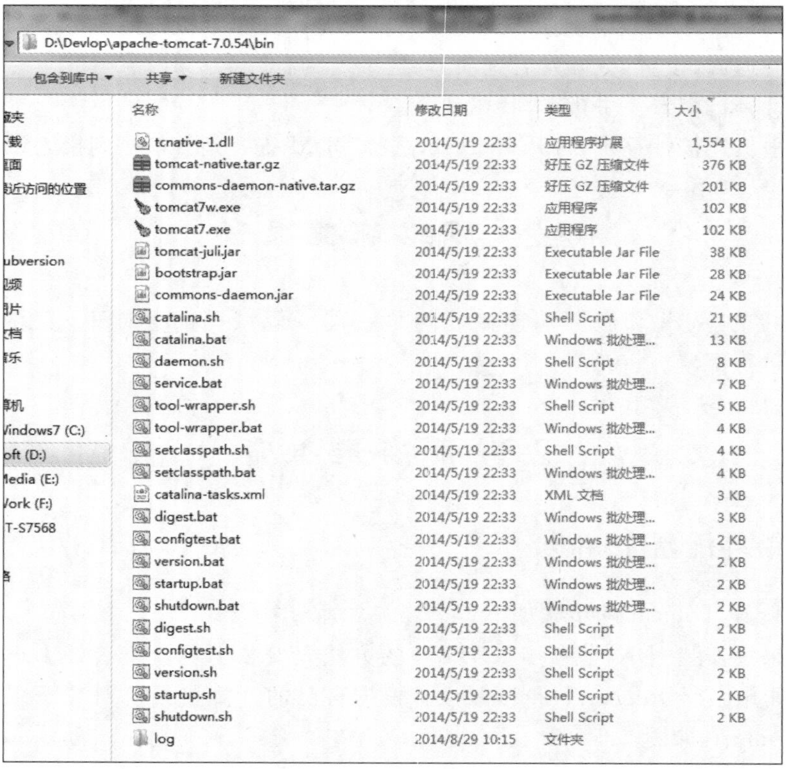

图 2.2　解压并进入 bin 文件夹

图 2.3 启动 Tomcat

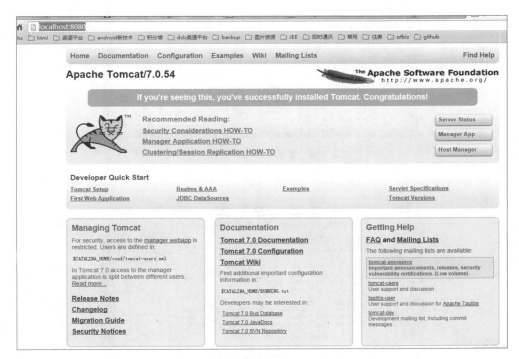

图 2.4 访问 http://localhost:8080/

2.3.2 安装并配置 MySQL

安装并配置 MySQL 的步骤如下：

（1）根据计算机系统版本下载 64 位或 32 位 MySQL，如图 2.5 所示。

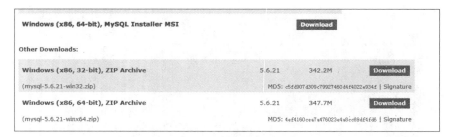

图 2.5 下载 MySQL

（2）安装 MySQL，单击 Next，如图 2.6 所示。

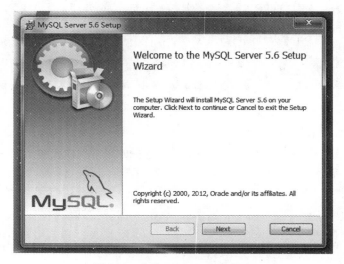

图 2.6 启动 MySQL 安装程序

（3）同意条款，单击 Next，如图 2.7 所示。

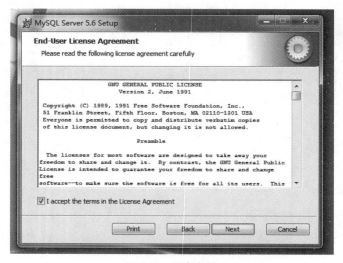

图 2.7 同意条款

(4) 选择 Custom,如图 2.8 所示。

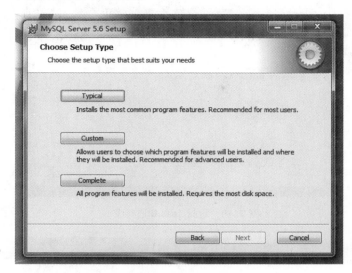

图 2.8 选择 Custom

(5) 选择所有组件,单击 Browse 选择安装路径,如图 2.9 所示。

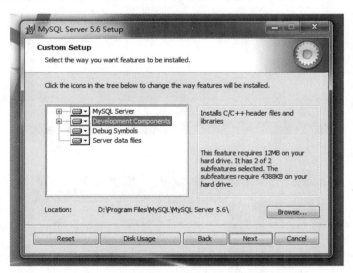

图 2.9 安装选项和安装路径

(6) 单击 Install 开始安装,如图 2.10、图 2.11、图 2.12 所示。

(7) 在 MySQL 的安装目录 bin 文件夹中,运行 MySQLInstanceConfig.exe 配置 MySQL,如图 2.13 所示。

(8) 单击 Next 开始安装,如图 2.14 所示。

(9) 选择 Detailed Configuration(详细配置),然后单击 Next 开始安装,如图 2.15 所示。

图 2.10 选择 Install 进行安装

图 2.11 等待安装

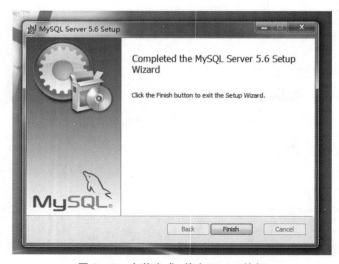

图 2.12 安装完成,单击 Finish 按钮

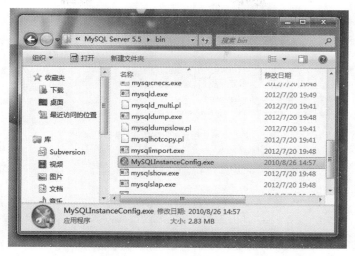

图 2.13 MySQLInstanceConfig.exe 文件

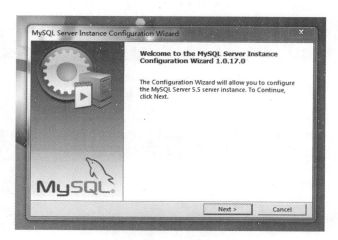

图 2.14 开始安装

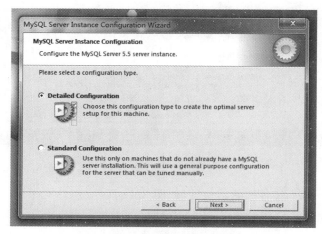

图 2.15 选择 Detailed Configuration

(10) 选择 Server Machine(占有资源、功能支持都一般,适合常规使用),然后单击 Next,如图 2.16 所示。

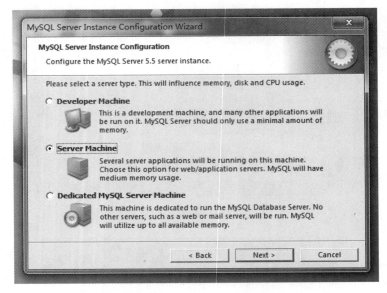

图 2.16 选择 Server Machine

(11) 选择 Transactional Database Only(只用来存储简单数据),如图 2.17 所示。

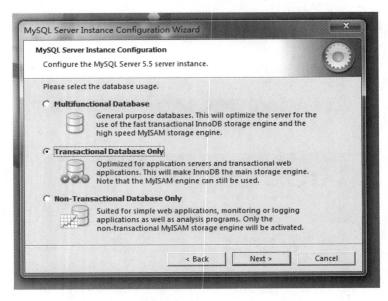

图 2.17 选择 Transactional Database Only

(12) 选择 MySQL 文件放置的位置,如图 2.18 所示。
(13) 选择 Manual Setting,如图 2.19 所示。
(14) 设置端口为 3306(重要,服务端程序设定的端口为 3306,如果不配置此端口,后

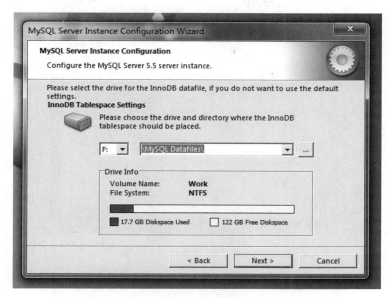

图 2.18 选择 MySQL 文件放置的位置

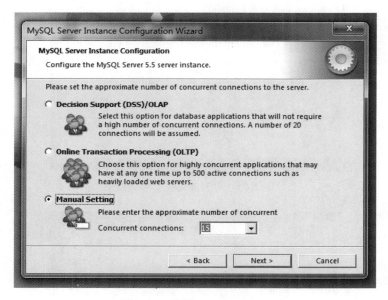

图 2.19 选择 Manual Setting

端程序访问数据库不会成功,所有的数据库操作都会失败),如图 2.20 所示。

(15) 设置编码类型为 utf8(重要,标准数据格式),如图 2.21 所示。

(16) 设置名字和勾选 PATH 相关选项,如图 2.22 所示。

(17) 设置 root 账户,密码为 root(重要,后端访问数据时,设定的数据库账户为 root,密码为 root,否则无法访问数据库),如图 2.23 所示。

(18) 执行配置选项,如图 2.24、图 2.25 所示。

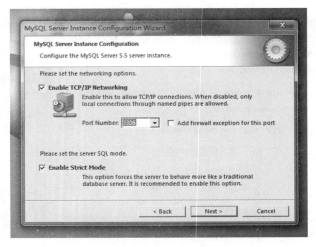

图 2.20　设置端口为 3306

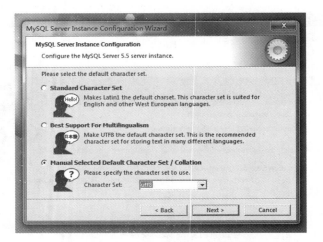

图 2.21　设置编码类型为 utf8

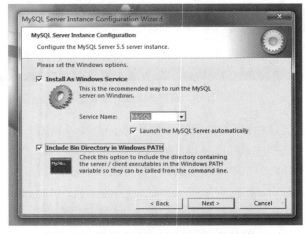

图 2.22　设置名字和勾选 PATH 相关选项

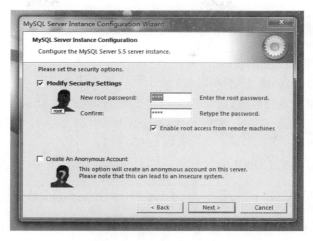

图 2.23 设置 root 账户

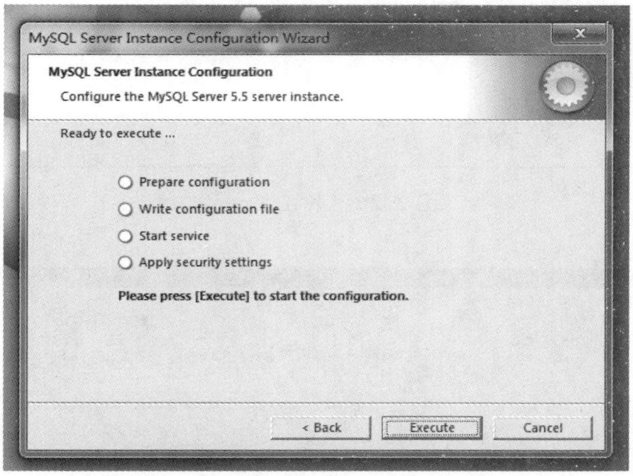

图 2.24 执行配置选项

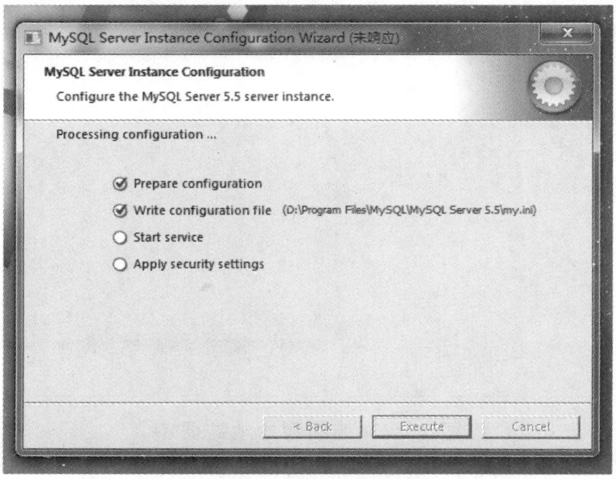

图 2.25 配置进行中

2.3.3　Navicat 的安装和使用

Navicat 的安装和使用步骤如下：

（1）下载 Navicat，如图 2.26 所示。

（2）Navicat 安装成功后，直接运行并新建连接，如图 2.27 所示。

（3）新建数据库 me_server 如图 2.28 所示，并在 me_server 数据库中新建表 me_user，如图 2.29 所示。

（4）建表成功，如图 2.30 所示。

图 2.26　下载 Navicat

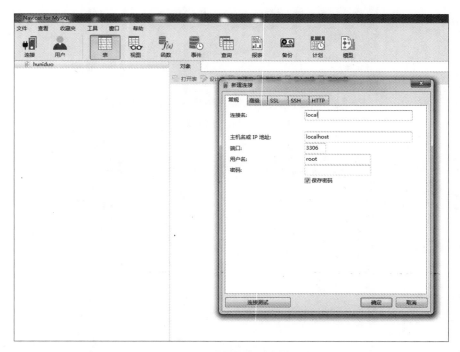

图 2.27　连接到本地数据库

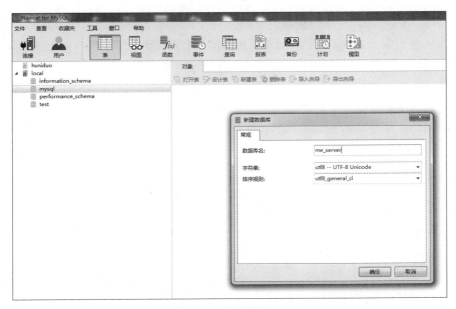

图 2.28 创建数据库

图 2.29 创建数据表

图 2.30 建表成功

2.4 知识点与技能回顾

本章主要知识点与技能如下：
（1）Java 开发的基础编程知识。
（2）SQLite 的增、删、改、查。
（3）开发过程的基本流程。

2.5 练 习

配置应用开发环境。
（1）在自己计算机上安装 Tomcat 并成功运行。
（2）将 meServer.zip 包解压到 Tomcat 的 webapps 目录中；运行 startup.bat＜Windows 环境下＞或 startup.sh＜Linux 环境下＞。
（3）在自己计算机上安装并配置 MySQL，用户名：root，密码：root＜密码不能空置，否则无法运行服务端程序＞。
（4）安装 Navicat，并建立 me_server 数据库，数据库端口设置为 3306，如图 2.27 所示。
（5）在 Navicat 上运行 SQL 脚本，建立三张表：me_user、me_shop、me_comment。
（6）访问注册接口：http://localhost：8080/meServer/user.php? act＝register&username＝test&password＝test；若得到结果如图 2.31 所示，则环境搭建成功。

图 2.31　环境搭建成功

第 3 章

MeDemo 项目介绍

3.1 项目背景

中国互联网络信息中心发布的第 41 次《中国互联网络发展状况统计报告》指出，截至 2017 年 12 月，我国网民规模达 7.72 亿，互联网普及率为 55.8%，较 2016 年年底提升 2.6 个百分点。我国手机网民规模达 7.53 亿，较 2016 年年底增加 5734 万人。网民中使用手机上网人群的占比由 2016 年的 95.1% 提升至 97.5%。

以上数据明确表明，移动互联网的发展已成为不可阻挡的趋势。随着移动互联网的兴起，各种移动营销服务乱象丛生，纷乱繁杂的应用令中小企业无所适从。本项目就是为了引导中小企业在移动营销中找准方向，避免盲目投资、重复投资，促进移动互联网处于良性发展的态势。

3.2 项目需求分析

在这个互联网高速发展的时代，越来越多的电子商务业务的出现，大大提高了人们生活的便利性。B2B、B2C、C2C 等各种业务形态如雨后春笋，移动互联网作为未来发展业务、拓展业务的利器，能够真正实现物联网的关键部分。

本软件是一款手机购物和商品管理系统，分别由手机客户端和服务端构成，可以查询商家与商品的各种信息。通过手机软件购物，可注册会员、积累积分。商家可以通过这个软件，添加、删除、修改、查询自己的商品，及时更新商品情况。

3.3 项目用例分析

后台用例分析如图 3.1 所示，分为管理用户、商户维护、评价管理、系统维护几部分。用户管理用例分析如图 3.2 所示，分为登录和未登录两种状态。

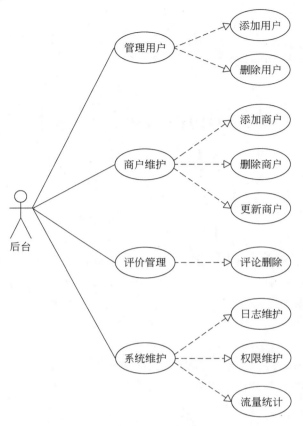

图 3.1 后台用例分析

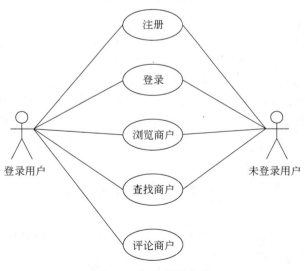

图 3.2 用户用例分析

3.4 项目流程

查找商户流程如图3.3所示。

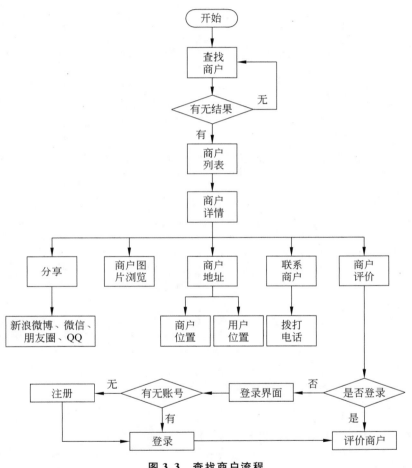

图3.3 查找商户流程

全部商户流程如图3.4所示。
首页流程如图3.5所示。

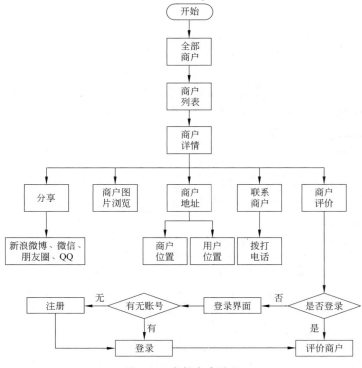

图 3.4　全部商户流程

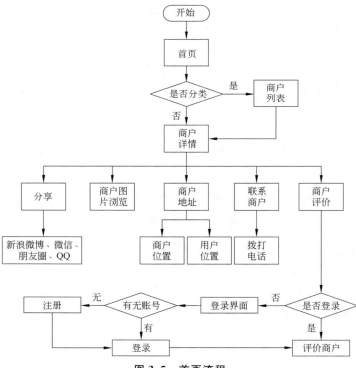

图 3.5　首页流程

3.5 项目数据库

项目数据库 E-R 图如图 3.6 所示。

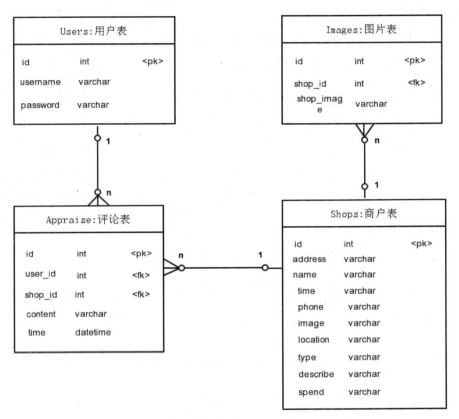

图 3.6 项目数据库 E-R 图

3.6 项目时序图

对于 Android 应用开发,整个系统最关键的地方就是与服务端的交互,包括注册、登录、查询商品、购买商品和添加商品等都是需要进行网络请求的操作。MeDemo 网络请求的时序图如图 3.7 所示。

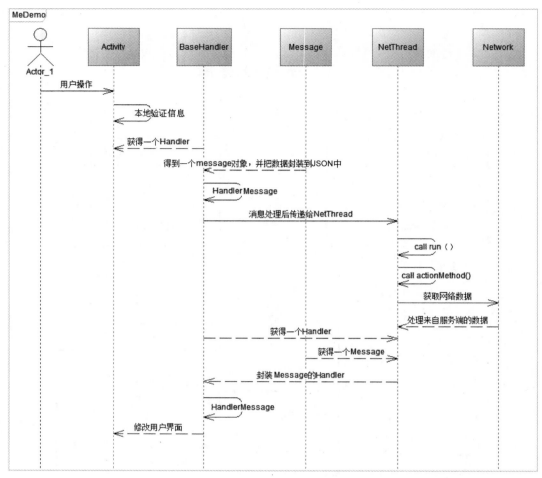

图 3.7 MeDemo 网络请求的时序图

第 4 章

用 户 注 册

4.1 用户注册总体设计

注册时序如图 4.1 所示。

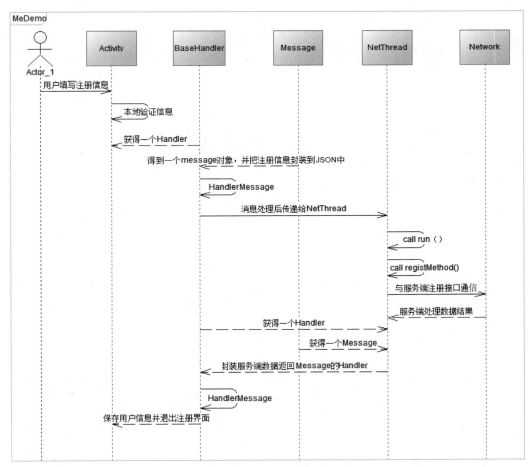

图 4.1 注册时序

注册流程如图 4.2 所示。

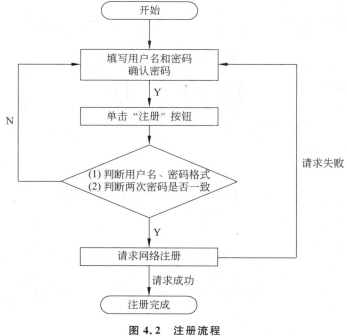

图 4.2　注册流程

4.2　数据库的准备

数据库用户属性如图 4.3 所示。

图 4.3　数据库用户属性

用户数据库表如表 4.1 所示。

表 4.1　用户数据库表

名　　称	说　　明	数据类型	主键/外键/非空
id	用户 id	int	P
username	用户名	varchar	
password	用户密码	varchar	

后台服务接口文档如下：

接口地址：http://localhost:8080/meServer/user.php。

调用方式：POST。

接口请求数据表如表4.2所示。

表 4.2 接口请求数据表

请求参数	必　　选	类型及范围	说　　明
act	Y	String	register
username	Y	String	用户名
password	Y	String	密码

返回方式：JSON。

调用示例：http://localhost:8080/meServer/user.php?act=register&username=321&password=321。

接口返回数据表如表4.3所示。

表 4.3 接口返回数据表

返回值字段	字 段 类 型	字 段 说 明
flag	Int	1：成功，0：失败
msg	String	消息提示信息
userInfo	JSONObject	用户信息Json数据
id	String	用户id
password	String	用户密码
username	String	用户名

4.3 用户注册重要知识点详解

4.3.1 Android项目目录结构

Android项目目录结构如图4.4所示。

其中：src是Java代码目录。libs是第三方库目录。drawable是xml资源文件目录。drawable_hdpi等是各尺寸的图片资源目录。values是字符串配置文件string.xml、尺寸配置文件dimen.xml、色值配置文件color.xml的目录。AndroidManifest.xml是项目的基本配置，包括项目版本、各类权限、4大组件注册等。proguard-project.txt是防止代码被反编译，完成代码混淆的配置文件。project.properties是编译以及引用其他库应用的配置文件。

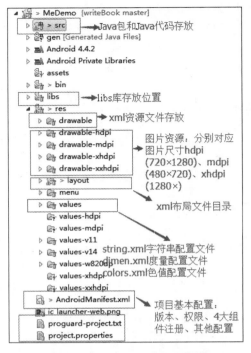

图 4.4　Android 项目目录结构

4.3.2　xml 布局文件的创建

1. 创建 xml 布局文件

xml 布局文件目录如图 4.5 所示。

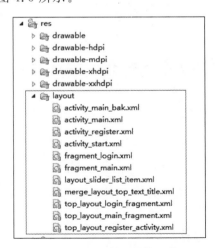

图 4.5　xml 布局文件目录

右键单击 layout 目录，新建 xml 布局文件如图 4.6 所示。

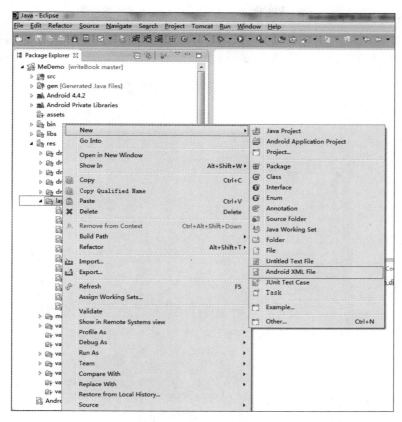

图 4.6 新建 xml 布局文件

输入 xml 布局文件名,可选择 LinearLayout、RelativeLayout 等布局方式,如图 4.7 所示。

图 4.7 输入布局文件名和选择布局方式

单击 Finish 按钮完成创建，如图 4.8 所示。

在 Layout 文件夹中，找到 activity_regist.xml 文件，如图 4.9 所示。双击鼠标左键，查看布局内容，如图 4.10 所示。

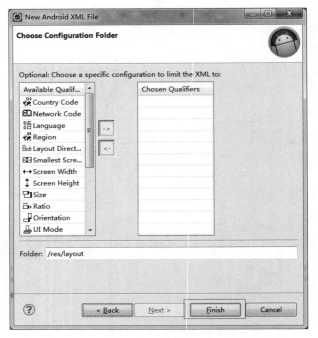

图 4.8　完成创建

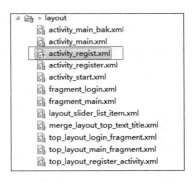

图 4.9　找到 activity_regist.xml 文件

```
1  <?xml version="1.0" encoding="utf-8"?>
2  <LinearLayout xmlns:android="http://schemas.android.com/apk/res/android"
3      android:layout_width="match_parent"
4      android:layout_height="match_parent"
5      android:orientation="vertical" >
6
7
8  </LinearLayout>
```

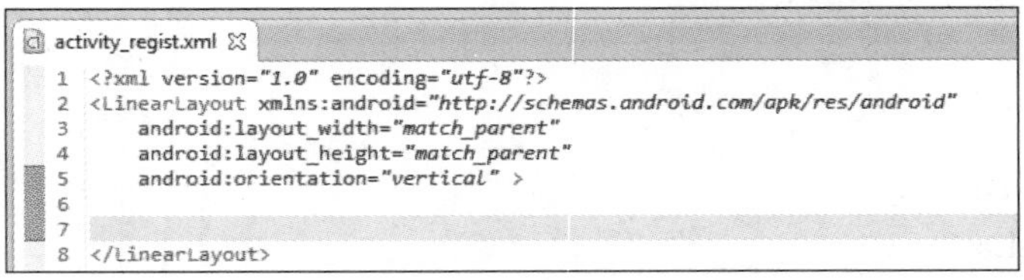

图 4.10　查看布局内容

2. 五大布局类型

1) LinearLayout

LinearLayout 按照垂直＜android：orientation＝" vertical "＞或者水平＜android：orientation＝"horizontal"＞的顺序依次排列子元素，每一个子元素都位于前一个元素之后。

如果是垂直排列，那么将是一个 N 行单列的结构，每一行只会有一个元素，而不论这个元素的宽度为多少；如果是水平排列，那么将是一个单行 N 列的结构。如果搭建两行两列的结构，通常的方式是先垂直排列两个元素，每一个元素里再包含一个 LinearLayout 进行水平排列。代码如下：

```xml
<?xml version="1.0" encoding="utf-8"?>
<LinearLayout xmlns:android="http://schemas.android.com/apk/res/android"
    android:layout_width="match_parent"
    android:layout_height="match_parent"
    android:orientation="vertical" >
    <TextView
        android:layout_width="wrap_content"
        android:layout_height="wrap_content"
        android:text="第 1 行" />
    <TextView
        android:layout_width="match_parent"
        android:layout_height="wrap_content"
        android:gravity="center_horizontal"
        android:text="第 2 行"/>
    <LinearLayout
        android:layout_width="match_parent"
        android:layout_height="wrap_content"
        android:orientation="horizontal" >
        <TextView
            android:layout_width="wrap_content"
            android:layout_height="wrap_content"
            android:text="第 1 列" />
        <TextView
            android:layout_width="wrap_content"
            android:layout_height="wrap_content"
            android:text="第 2 列" />
    </LinearLayout>
</LinearLayout>
```

运行效果如图 4.11 所示。

2) RelativeLayout

RelativeLayout 按照各子元素之间的位置关系完成布局。在此布局中的子元素里与位置相关的属性将生效，例如 android：layout_below、android：layout_above 等。子元素就通过这些属性和各自的 ID 配合指定位置关系。注意在指定位置关系时,引用的 ID 必须在引用之前先被定义,否则将出现异常。

下面介绍 RelativeLayout 中常用的位置属性。

android:layout_toLeftOf:该组件位于引用组件的左方。

android:layout_toRightOf:该组件位于引用组件的右方。

android:layout_above:该组件位于引用组件的上方。

android:layout_below:该组件位于引用组件的下方。

android:layout_alignParentLeft:该组件是否对齐父组件的左端。

android:layout_alignParentRight:该组件是否对齐父组件的右端。

android:layout_alignParentTop:该组件是否对齐父组件的顶部。

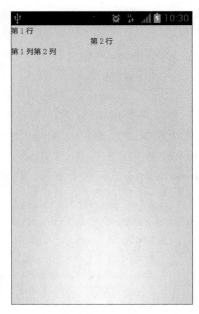

图 4.11　垂直与水平布局运行效果

android:layout_alignParentBottom:该组件是否对齐父组件的底部。

android:layout_centerInParent:该组件是否相对于父组件居中。

android:layout_centerHorizontal:该组件是否横向居中。

android:layout_centerVertical:该组件是否垂直居中。

RelativeLayout 是 Android 五大布局结构中最灵活的一种布局结构,比较适合复杂界面的布局。代码如下:

```
<?xml version="1.0" encoding="utf-8"?>
<RelativeLayout xmlns:android="http://schemas.android.com/apk/res/android"
    android:layout_width="match_parent"
    android:layout_height="match_parent" >
    <TextView
        android:id="@+id/text_01"
        android:layout_width="50dp"
        android:layout_height="50dp"
        android:layout_alignParentBottom="true"
        android:background="#ffffffff"
        android:gravity="center"
        android:text="底部 1" />
    <TextView
        android:id="@+id/text_02"
        android:layout_width="100dp"
        android:layout_height="100dp"
```

```
                android:layout_above="@id/text_01"
                android:layout_centerHorizontal="true"
                android:background="@android:color/holo_blue_light"
                android:gravity="center"
                android:text="底部 2\n 在底部 1 的上面" />
        <TextView
                android:id="@+id/text_03"
                android:layout_width="100dp"
                android:layout_height="50dp"
                android:layout_above="@id/text_01"
                android:layout_toLeftOf="@id/text_02"
                android:background="#fffedcba"
                android:gravity="center"
                android:text="底部 3\n 在底部 1 的上面,底部 2 的左边" />
</RelativeLayout>
```

运行效果如图 4.12 所示。

图 4.12　RelativeLayout 布局运行效果

3) TableLayout

TableLayout 为表格布局,适用于 N 行 N 列的布局格式。一个 TableLayout 由许多 TableRow 组成,一个 TableRow 就代表 TableLayout 中的一行。

TableRow 是 LinearLayout 的子类,它的 android：orientation 属性值恒为 horizontal,并且它的 android：layout_width 和 android：layout_height 属性值恒为 MATCH_PARENT 和 WRAP_CONTENT,所以它的子元素都是横向排列,并且宽高一致。这样的设计使得每个 TableRow 里的子元素都相当于表格中的单元格。在

TableRow 中，单元格可以为空，但是不能跨列。代码如下：

```xml
<?xml version="1.0" encoding="utf-8"?>
<TableLayout xmlns:android="http://schemas.android.com/apk/res/android"
    android:layout_width="match_parent"
    android:layout_height="match_parent"
    android:gravity="center" >
    <TableRow
        android:layout_width="wrap_content"
        android:layout_height="wrap_content"
        android:gravity="center" >
        <TextView
            android:background="@android:color/holo_blue_light"
            android:gravity="center"
            android:padding="10dp"
            android:text="1行1列" />
        <TextView
            android:background="@android:color/holo_green_light"
            android:gravity="center"
            android:padding="10dp"
            android:text="1行2列" />
        <TextView
            android:background="@android:color/holo_orange_light"
            android:gravity="center"
            android:padding="10dp"
            android:text="1行3列" />
    </TableRow>
    <TableRow
        android:layout_width="wrap_content"
        android:layout_height="wrap_content"
        android:gravity="center" >
        <TextView
            android:background="@android:color/holo_green_light"
            android:gravity="center"
            android:padding="10dp"
            android:text="2行1列" />
        <TextView
            android:background="@android:color/holo_orange_light"
            android:gravity="center"
            android:padding="10dp"
            android:text="2行2列" />
    </TableRow>
    <TableRow
        android:layout_width="wrap_content"
        android:layout_height="wrap_content"
        android:gravity="center" >
```

```
        <TextView
            android:background="@android:color/holo_orange_light"
            android:gravity="center"
            android:padding="10dp"
            android:text="3行1列" />
        <TextView
            android:background="@android:color/holo_green_light"
            android:gravity="center"
            android:padding="10dp"
            android:text="3行2列" />
        <TextView
            android:background="@android:color/holo_blue_light"
            android:gravity="center"
            android:padding="10dp"
            android:text="3行3列" />
    </TableRow>
</TableLayout>
```

布局效果如图4.13所示。

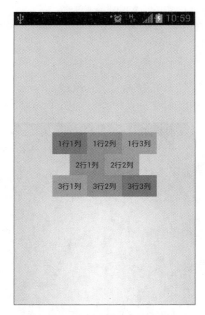

图4.13　TableLayout布局效果

4）FrameLayout

FrameLayout是五大布局中最简单的一个布局。在这个布局中,整个界面被当成一块空白备用区域,所有的子元素都不能被指定放置的位置,它们全部放于这块区域的左上角,并且后面的子元素直接覆盖在前面的子元素上,将前面的子元素部分和全部遮挡。

```
<?xml version="1.0" encoding="utf-8"?>
```

```
<FrameLayout xmlns:android="http://schemas.android.com/apk/res/android"
    android:layout_width="fill_parent"
    android:layout_height="fill_parent"
    android:orientation="vertical" >
    <TextView
        android:layout_width="fill_parent"
        android:layout_height="fill_parent"
        android:background="@android:color/holo_green_light"
        android:gravity="center"
        android:text="第 1 块区域" />
    <TextView
        android:layout_width="120dp"
        android:layout_height="120dp"
        android:background="@android:color/holo_blue_light"
        android:gravity="center"
        android:text="第 2 块区域" />
    <TextView
        android:layout_width="50dp"
        android:layout_height="50dp"
        android:background="@android:color/holo_orange_light"
        android:gravity="center"
        android:text="第 3 块区域" />
</FrameLayout>
```

运行效果如图 4.14 所示。

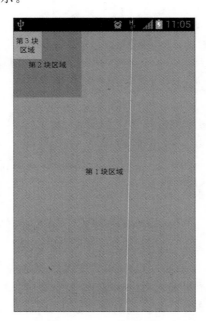

图 4.14　FrameLayout 布局运行效果

5）AbsoluteLayout

AbsoluteLayout 是绝对位置布局。在此布局中的子元素的 android：layout_x 和 android：layout_y 属性将生效，用于描述该子元素的坐标位置。屏幕左上角为坐标原点(0,0)。第一个 0 代表横坐标，向右移动此值增大；第二个 0 代表纵坐标，向下移动此值增大。在此布局中的子元素可以相互重叠。在实际开发中，通常不采用此布局格式，因为它的界面代码过于刚性，以至于有可能不能很好地适配各种终端。代码如下：

```
<AbsoluteLayout xmlns:android="http://schemas.android.com/apk/res/android"
    android:layout_width="fill_parent"
    android:layout_height="fill_parent"
    android:orientation="vertical" >
    <TextView
        android:layout_width="50dp"
        android:layout_height="50dp"
        android:layout_x="50dp"
        android:layout_y="50dp"
        android:background="@android:color/holo_orange_light"
        android:gravity="center"
        android:text="第 1 个布局" />
    <TextView
        android:layout_width="50dp"
        android:layout_height="50dp"
        android:layout_x="25dp"
        android:layout_y="25dp"
        android:background="@android:color/holo_blue_light"
        android:gravity="center"
        android:text="第 2 个布局" />
    <TextView
        android:layout_width="50dp"
        android:layout_height="50dp"
        android:layout_x="125dp"
        android:layout_y="125dp"
        android:background="@android:color/holo_green_light"
        android:gravity="center"
        android:text="第 3 个布局" />
</AbsoluteLayout>
```

布局效果如图 4.15 所示。

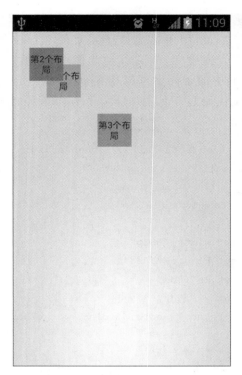

图 4.15　AbsoluteLayout 布局效果

4.3.3　Activity 的创建

1. 如何创建 Activity

src 目录结构如图 4.16 所示。

鼠标右键单击 src 目录,创建 Package 如图 4.17 所示。

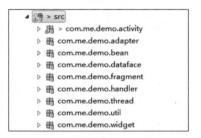

图 4.16　src 目录结构

输入 Package 的名称,如图 4.18 所示。

src 目录结构变化,新建的 test 包如图 4.19 所示。

右键单击新建的 test 包,选择 New,并单击 Class 按钮,创建 Class 文件,如图 4.20 所示。

输入 TestActivity 名称等,并继承 Activity,然后单击 Finish 按钮,如图 4.21 所示。

查看创建成功的 TestActivity,如图 4.22 所示。

在 TestActivity 中,重写 Activity 的 onCreate(Bundle SaveInstanceState)方法,并设置 setContentView(int layoutId)加载 xml 布局文件,如图 4.23 所示。

第 4 章 用户注册

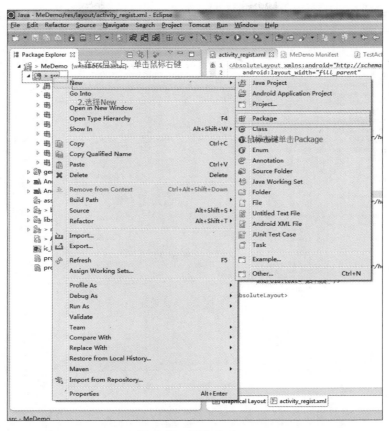

图 4.17 创建 Package

图 4.18 输入 Package 的名称

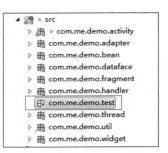

图 4.19 新建的 test 包

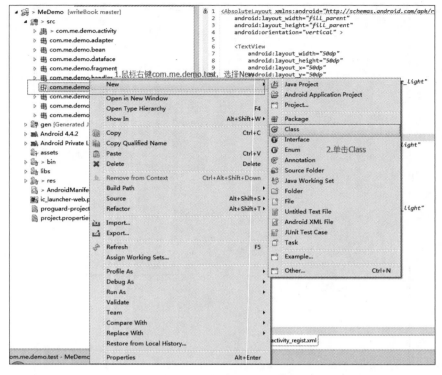

图 4.20 创建 Class 文件

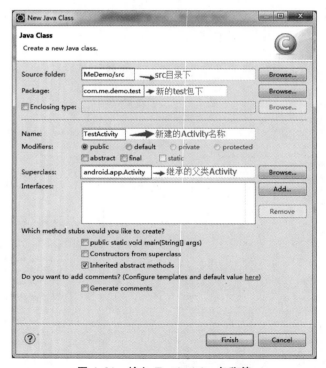

图 4.21 输入 TestActivity 名称等

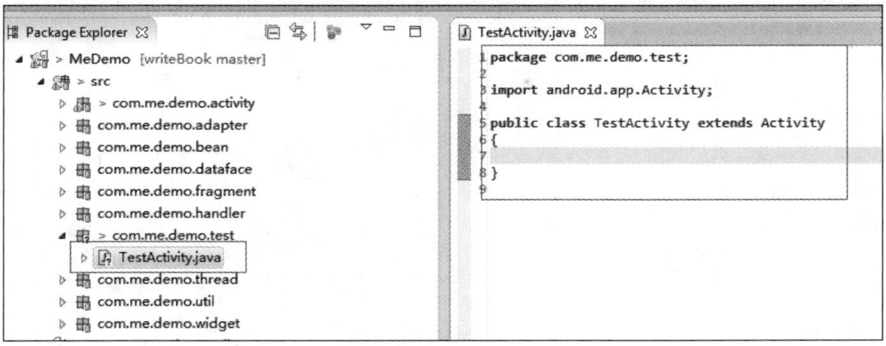

图 4.22　创建成功后的 TestActivity

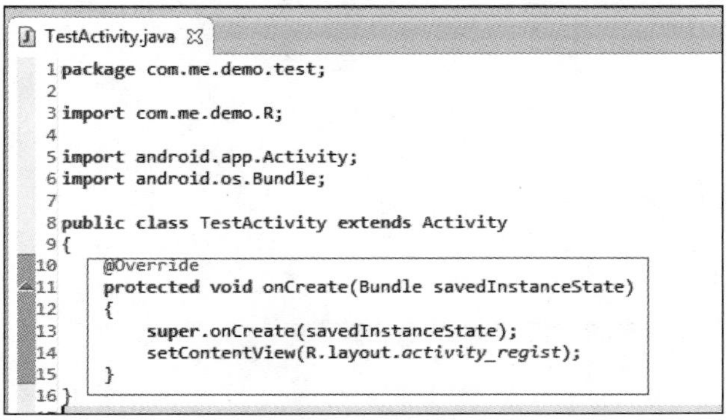

图 4.23　重写 onCreate 方法并加载布局文件

布局文件代码如下所示。

```xml
<?xml version="1.0" encoding="utf-8"?>
<LinearLayout xmlns:android="http://schemas.android.com/apk/res/android"
    android:layout_width="match_parent"
    android:layout_height="match_parent"
    android:gravity="center"
    android:orientation="vertical" >
    <TextView
        android:layout_width="wrap_content"
        android:layout_height="wrap_content"
        android:text="这是我的第一个 Android 应用"
        android:textSize="22sp"
        android:textStyle="bold" />
</LinearLayout>
```

在 Manifest.xml 中注册 TestActivity 并且设置为程序入口 Activity，Manifest.xml 结构如图 4.24 所示。

图 4.24　Manifest.xml 结构

2. 运行 Android 程序

选择 DDMS 如图 4.25 所示。

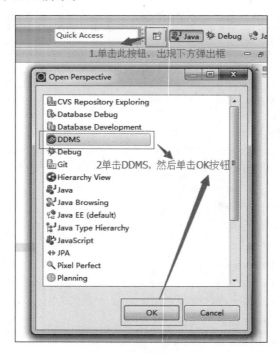

图 4.25　选择 DDMS

在 DDMS 中的 Devices 里查看手机是否已连接到 Eclipse，如图 4.26 所示。单击右键，选择 Run As。单击 Android Application，如图 4.27 所示。

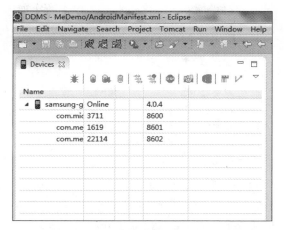

图 4.26 手机已经成功连接到 Eclipse

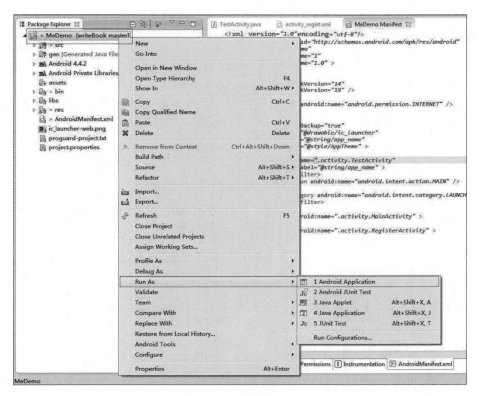

图 4.27 运行程序

4.3.4 dimen 资源文件

字体大小设置 android：textSize＝"@dimen/text_size_18"，顶部 actionBar 高度设置 android：layout_height＝"@dimen/action_bar_default_height"，都用到了 dimen 属性资源。dimen 一般配置如下：

```xml
<resources>
    <dimen name="activity_horizontal_margin">16dp</dimen>
    <dimen name="activity_vertical_margin">16dp</dimen>
    <dimen name="action_bar_default_height">50dp</dimen>
    <dimen name="text_size_20">20sp</dimen>
    <dimen name="text_size_18">18sp</dimen>
    <dimen name="text_size_16">16sp</dimen>
    <dimen name="text_size_14">14sp</dimen>
</resources>
```

配置的使用有助于资源的复用，同时也便于修改。Android 有各种分辨率、各种尺寸，因此界面布局时，需要考虑与手机适配。Google 有几种处理方式，其中 dimen 文件配置是其中之一。如同 drawable 一样，可以建立 ldpi、mdpi、hdpi、xhdpi、xxhdpi 等几种适配文件资源。res 资源目录结构如图 4.28 所示。

4.3.5　drawable 资源文件

返回按钮是 ImageView，图片设置 android：src＝"@drawable/icon_back"。注册按钮是 Button，背景设置 android：background＝"@drawable/bkg_button"。一般 drawable 对应 dpi 里面存放的相应的图片资源。这里用的不是图片资源，而是 drawable 里的 xml 文件，如图 4.29 所示。

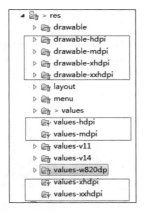
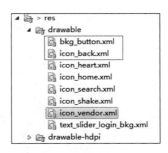

图 4.28　res 资源目录结构　　　图 4.29　文件所在位置

bkg_button.xml 内容如下：

```xml
<?xml version="1.0" encoding="utf-8"?>
<selector xmlns:android="http://schemas.android.com/apk/res/android" >
    <item android:drawable="@drawable/funcation_button_pressed" android:state_pressed="true"/>
    <item android:drawable="@drawable/funcation_button_normal"/>
</selector>
```

icon_back.xml 内容如下：

```xml
<?xml version="1.0" encoding="utf-8"?>
<selector xmlns:android="http://schemas.android.com/apk/res/android" >
    <item android:drawable="@drawable/back_button_pressed" android:state_pressed="true"/>
    <item android:drawable="@drawable/back_button_normal"/>
</selector>
```

符合某种场景时,调用对应的图片资源,提供更好的交互体验。例如,返回按钮未被按下,是 back_button_normal.png 图片,但是当图片被按下时,按钮背景变为 back_button_pressed.png 图片。

selector 属性介绍如下:

android:state_selected:选中。

android:state_focused:获得焦点。

android:state_pressed:单击。

android:state_enabled:设置是否响应事件,指所有事件。

4.3.6 客户端与服务器的交互

将数据封装到 JSON 对象中,代码如下。

```
param.put("act", "register");
param.put("username", userName);
param.put("password", password);
```

服务端返回 JSON String,解析放到 JavaBean 文件 UserBean.java 中。代码如下:

```java
package com.me.demo.bean;

public class UserBean
{
    public String id;
    public String username;
    public String password;}
```

4.4 用户注册

4.4.1 用户注册的具体实现

1. 注册界面效果及实现

注册界面如图 4.30 所示。

1) 顶部组成

(1) 返回按钮(ImageView)。

ImageView 是加载图片的组件。

android：id 设置组件的 id，生成该 View 的唯一标识符。

android：layout_width 设置该组件的宽度，设置为 dimen 中定义的宽度＜参照 4.4.3 节＞。

android：layout_height 设置该组件的高度。wrap_content 代表根据这个组件加载内容自适应，可以设置为 match_parent，表示占满父空间的空间；也可以设置固定的值。这里由于我们的图标比较小，因此高度选择的是 wrap_content。

android：layout_alignParentRight 设置该组件在父 View 的最右侧。只有父组件为 RelativeLayout 时，才有这个属性。

android：layout_alignParentLeft 设置该组件在父 View 的最左侧。只有父组件为 RelativeLayout 时，才有这个属性。

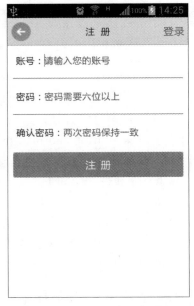

图 4.30　用户注册界面效果图

android：padding 设置该组件内部与该组件边距为 8dip。

android：src 是 ImageView 最主要的属性，设置显示的图片。@null 表示不显示任何图片，@drawable/icon_back 显示在 Drawable 中资源文件名为 icon_back 的图片，或 icon_back 的 xml 文件。

（2）标题（TextView）。

TextView 包含在 include 内。由于 TextView 的父类也是 RelativeLayout 布局，所以我们采用 merge 方式。

merge 对于优化布局冗余非常重要。如果布局 Layout 和父类 Layout 是同一类型，则可以将 merge 嵌入到父类中，并且可以减少视图层级。

android：layout_centerInParent＝"true" 与父控件的关系是居中。当父控件为 RelativeLayout 时，才有这个属性。

android：gravity＝"center_horizontal"设置控件字体布局为水平居中。

android：text＝"@string/register"设置显示的字体为定义的字符串，如图 4.31 所示。

android：textSize＝"@dimen/text_size_18"设置显示的字体大小，如图 4.32 所示。

代码如下：

```
<?xml version="1.0" encoding="utf-8"?>
<merge xmlns:android="http://schemas.android.com/apk/res/android"
    xmlns:tools="http://schemas.android.com/tools"
    android:layout_width="match_parent"
    android:layout_height="match_parent" >
    <TextView
```

```
        android:id="@+id/text_top_title"
        android:layout_width="match_parent"
        android:layout_height="wrap_content"
        android:layout_centerInParent="true"
        android:gravity="center_horizontal"
        android:text="@string/register"
        android:textSize="@dimen/text_size_18" />
</merge>
```

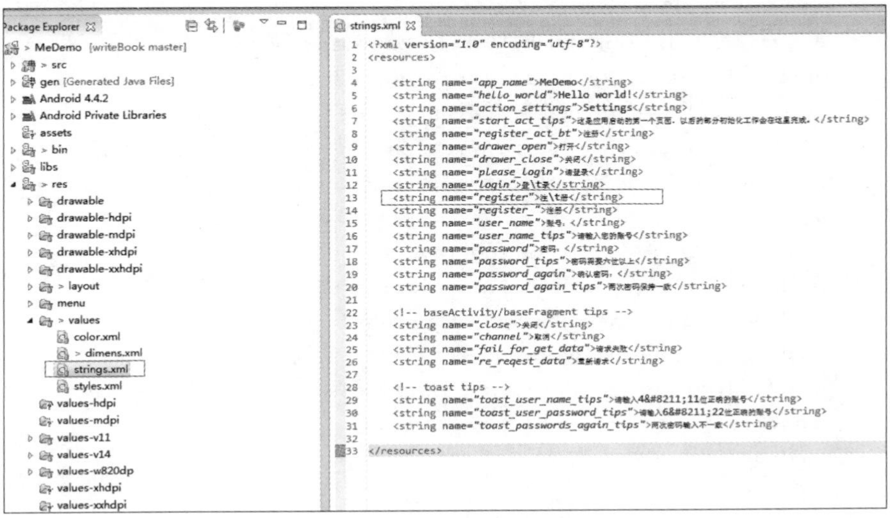

图 4.31　strings.xml

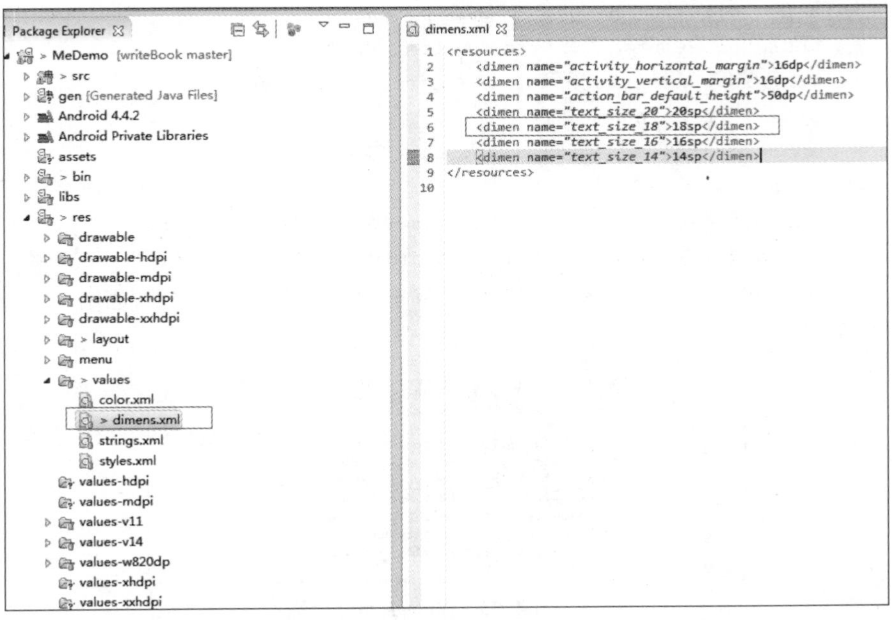

图 4.32　dimens.xml

(3) 父视图(RelativeLayout)。

android：layout_width="match_parent"设置宽度全屏。

android：layout_height="@dimen/action_bar_default_height"对不同的手机配置不同的dimens.xml，来设定对应的不同高度，如图4.33所示。

android：background="@drawable/top_layout_supplier_action_bar"设置背景为图4.34所示的资源图片，适配不同的手机，可以制作针对高、中、低分辨率的图片。代码如下：

```xml
<?xml version="1.0" encoding="utf-8"?>
<RelativeLayout xmlns:android="http://schemas.android.com/apk/res/android"
    xmlns:tools="http://schemas.android.com/tools"
    android:layout_width="match_parent"
    android:layout_height="@dimen/action_bar_default_height"
    android:background="@drawable/top_layout_supplier_action_bar" >
    <ImageView
        android:id="@+id/image_top_layout_right"
        android:layout_width="@dimen/action_bar_default_height"
        android:layout_height="match_parent"
        android:layout_alignParentRight="true"
        android:contentDescription="@string/app_name"
        android:padding="8dip"
        android:src="@null" />
    <ImageView
        android:id="@+id/image_top_layout_left"
        android:layout_width="@dimen/action_bar_default_height"
        android:layout_height="match_parent"
        android:layout_alignParentLeft="true"
        android:contentDescription="@string/app_name"
        android:padding="8dip"
        android:src="@drawable/icon_back" />
    <include
        android:layout_toLeftOf="@id/image_top_layout_right"
        android:layout_toRightOf="@id/image_top_layout_left"
        layout="@layout/merge_layout_top_text_title" />
</RelativeLayout>
```

```
dimens.xml
1  <resources>
2      <dimen name="activity_horizontal_margin">16dp</dimen>
3      <dimen name="activity_vertical_margin">16dp</dimen>
4      <dimen name="action_bar_default_height">50dp</dimen>
5      <dimen name="text_size_20">20sp</dimen>
6      <dimen name="text_size_18">18sp</dimen>
7      <dimen name="text_size_16">16sp</dimen>
8      <dimen name="text_size_14">14sp</dimen>
9  </resources>
```

图4.33　dimens.xml

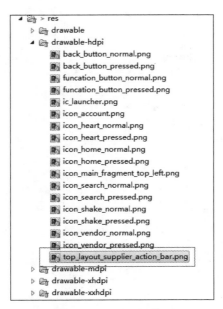

图 4.34　drawable 图片资源

2) 中间信息组成

中间信息包括信息名(TextView)、信息输入框(EditText)、父视图(LinearLayout)。

(1) 信息输入框(EditText)。

android：id="@+id/edit_username"设置组件 id,后续使用该组件时需要它。

android：layout_width="match_parent"设置宽度全屏。

android：layout_height 设置该组件的高度自适应。

android：orientation="horizontal"设置为水平布局。

android：padding="5dp"该组件内部边距设置为 5dp,表示内部的文字与边框距离 5dp。

android：layout_margin="10dp"设置该组件四边的外部距离为 10dp,表示该组件与其他组件或者父组件的边框至少距离 10dp。

部分代码如下。

```
<?xml version="1.0" encoding="utf-8"?>
<LinearLayout xmlns:android="http://schemas.android.com/apk/res/android"
    xmlns:tools="http://schemas.android.com/tools"
    android:layout_width="match_parent"
    android:layout_height="match_parent"
    android:background="@color/white"
    android:orientation="vertical"
    tools:context="com.me.demo.activity.RegisterActivity" >
    <include
        android:layout_width="match_parent"
```

```xml
        android:layout_height="@dimen/action_bar_default_height"
        layout="@layout/top_layout_register_activity" />
    <LinearLayout
        android:layout_width="match_parent"
        android:layout_height="wrap_content"
        android:layout_margin="10dp"
        android:orientation="horizontal"
        android:padding="5dp" >
        <TextView
            android:layout_width="wrap_content"
            android:layout_height="wrap_content"
            android:text="@string/user_name"
            android:textSize="@dimen/text_size_18" />
        <EditText
            android:id="@+id/edit_username"
            android:layout_width="match_parent"
            android:layout_height="wrap_content"
            android:background="@null"
            android:hint="@string/user_name_tips" />
    </LinearLayout>
    <View
        android:layout_width="match_parent"
        android:layout_height="1dp"
        android:layout_marginBottom="5dp"
        android:layout_marginLeft="10dp"
        android:layout_marginRight="10dp"
        android:layout_marginTop="5dp"
        android:background="@android:color/darker_gray" />
    <LinearLayout
        android:layout_width="match_parent"
        android:layout_height="wrap_content"
        android:layout_margin="10dp"
        android:orientation="horizontal"
        android:padding="5dp" >
        <TextView
            android:layout_width="wrap_content"
            android:layout_height="wrap_content"
            android:text="@string/password"
            android:textSize="@dimen/text_size_18" />
        <EditText
            android:id="@+id/edit_password"
            android:layout_width="match_parent"
            android:layout_height="wrap_content"
            android:background="@null"
```

```xml
            android:hint="@string/password_tips"
            android:inputType="textPassword" >
        </EditText>
    </LinearLayout>
    <View
        android:layout_width="match_parent"
        android:layout_height="1dp"
        android:layout_marginBottom="5dp"
        android:layout_marginLeft="10dp"
        android:layout_marginRight="10dp"
        android:layout_marginTop="5dp"
        android:background="@android:color/darker_gray" />
    <LinearLayout
        android:layout_width="match_parent"
        android:layout_height="wrap_content"
        android:layout_margin="10dp"
        android:orientation="horizontal"
        android:padding="5dp" >
        <TextView
            android:layout_width="wrap_content"
            android:layout_height="wrap_content"
            android:text="@string/password_again"
            android:textSize="@dimen/text_size_18" />
        <EditText
            android:id="@+id/edit_password_again"
            android:layout_width="match_parent"
            android:layout_height="wrap_content"
            android:background="@null"
            android:hint="@string/password_again_tips"
            android:inputType="textPassword" />
    </LinearLayout>
    <Button
        android:id="@+id/button_register"
        android:layout_width="match_parent"
        android:layout_height="wrap_content"
        android:layout_margin="10dp"
        android:background="@drawable/bkg_button"
        android:padding="10dp"
        android:text="@string/register"
        android:textColor="@color/white"
        android:textSize="@dimen/text_size_20" />
</LinearLayout>
```

（2）注册按钮（Button）。

android：layout_margin＝"10dp"设置外部四边的距离为 10dp。

android：layout_marginLeft 设置与左边控件的距离，android：layout_marginRight 设置与右边控件的距离。同理设置与上方和下方控件的距离。

android：padding 设置内部四边的距离，与 margin 相似，也有上、下、左、右 4 个方向。

android：textColor＝"@color/white"设置字体颜色，在 color.xml 配置文件中，如图 4.35 所示。代码如下：

```
<Button
        android:layout_width="match_parent"
        android:layout_height="wrap_content"
        android:layout_margin="10dp"
        android:background="@drawable/bkg_button"
        android:padding="10dp"
        android:text="@string/register"
        android:textColor="@color/white"
        android:textSize="@dimen/text_size_20" />
```

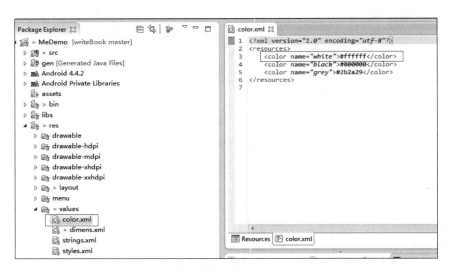

图 4.35　color.xml

2. 注册流程控制

1）RegisterActivity 获取 activity_register 的数据信息

（EditText）findViewById(R.id.edit_username)是通过在 xml 设置的 id 找到对应的组件，然后强制转换为 id 对应的组件。

userNameEdit.getText()得到 Editable，连接空字符串""即可得到字符串。

用户名：获得输入的用户名字符串。

```
EditText userNameEdit =(EditText) findViewById(R.id.edit_username);
```

```
userName =userNameEdit.getText() +"";
```

密码：获得输入的密码字符串。

```
EditText passwordEdit = (EditText) findViewById(R.id.edit_password);
password =passwordEdit.getText() +"";
```

2）检查信息填写是否符合规范

trim()消除空格影响，通过正则表达式检查是否符合要求。

检查用户名：用户名由3～11位大小写字母或者数字组成。不符合此条件时，提示用户名不符合规则。

String.matches()方法用Java的String类判断String对象是否符合正则表达式的规则。

WidgetTools类的setTVError方法弹出信息提示框，提示用户输入是否符合规则。

```
if (!userName.trim().matches("^[a-zA-Z0-9_][a-zA-Z0-9_]{3,10}$"))
{
    WidgetTools.setTVError (userNameEdit, this.getResources().getString(R.string.toast_user_name_tips), this);
    return;
}
```

检查密码：密码组成规则，由大小写字母、数字或者部分特殊字符组成。不符合此条件则弹出错误提示框，并提示密码不符合规则。

```
if (!password.trim().matches("^[\\@A-Za-z0-9\\!\\#\\$\\%\\^\\&\\*\\.\\~]{5,22}$"))
{
    WidgetTools.setTVError(passwordEdit, getResources().getString(R.string.toast_user_password_tips), this);
    return;
}
```

检查重复密码：确认与第一次输入的密码是否一致。

```
if (!passwordAgain.equals(password))
{
    WidgetTools.setTVError (passwordAgainEdit, getResources().getString(R.string.toast_passwords_again_tips), this);
    return;
}
```

正则表达式根据特定规则检测字符串组成的规则，在此不详细介绍。在第10章中，会详细介绍正则表达式的知识。

3）检查通过，封装数据到JSON中

sendMessage方法是自定义的BaseActivity实现DataCallBack接口的方法。此处重

写这个方法。

数据传递到服务端,数据格式是 JSON 格式。JSONObject 是 JSON 的一个对象类,它以键值对的形式存在。JSONObject 的 put 方法,将键值对保存到 JSONObject 对象中,传递给后端模块处理。

关于 JSON 的知识点,本章不再详细讲解,在第 5 章的知识点扩展中会详细介绍。

请求服务端数据是一个互通的过程,需要时间去处理。为使用户有良好的体验,需要在界面显示当前正在请求网络,请用户耐心等待。Android 的一般处理方式是在当前页面弹出 ProgressDialog。利用 ProgressaDialog 对象的 show 方法,就会显示一个 Dialog 在页面上。代码如下:

```java
@Override
public void sendMessage(int opt)
{
    progressDialog.show();
    message =mHandler.obtainMessage();
    param =new JSONObject();
    try
    {
        switch (opt)
        {
            case OPT.USER_REGISTER:
                param.put("act", "register");
                param.put("username", userName);
                param.put("password", password);
                break;
        }
    } catch (JSONException e)
    {
        e.printStackTrace();
    }
    super.sendMessage(opt);
}
```

4)通过 Handler 传递到 BaseHandler 的 handleMessage

sendMessage 方法将数据封装完毕后,发送信息到该 Activity 的父类的 sendMessage 方法。

Android 系统的信息传递机制,由 Message、Handler 和 Looper 组成。

(1) Message:Message 在 android.os 包中,定义一个 Message 包含的必要的描述和属性数据,并且此对象可以被发送给 android.os.Handler 处理。属性字段包括 arg1、arg2、what、obj、replyTo 等。其中 arg1 和 arg2 是用来存放整型数据的;what 是用来保存消息标示的;obj 是 Object 类型的任意对象;replyTo 是消息管理器,会关联到一个 handler,handler 处理其中的消息。

通常对 Message 对象不直接使用 new 来创建，而是调用 handler 中的 obtainMessage 方法来直接获得 Message 对象。

（2）Handler：Handler 主要有两个用途：首先是可以定时处理或者分发消息，其次是可以添加一个执行的行为在其他线程中执行。

对于 Handler 中的方法，可以选择你关心的操作去覆盖它，处理具体的业务操作，常见的就是对消息的处理可以覆盖 public void handleMessage（参数）方法，可以根据参数选择对此消息是否需要做出处理，这个和具体的参数有关。

（3）Looper：Looper 主要管理消息队列，负责消息的出列和入列操作。Message 的 obj 属性保存 Activity 中封装的 JSONObject 对象 param。Message 的 what 属性保存当前请求接口的 opt 常量。

接口实现如图 4.36 所示。

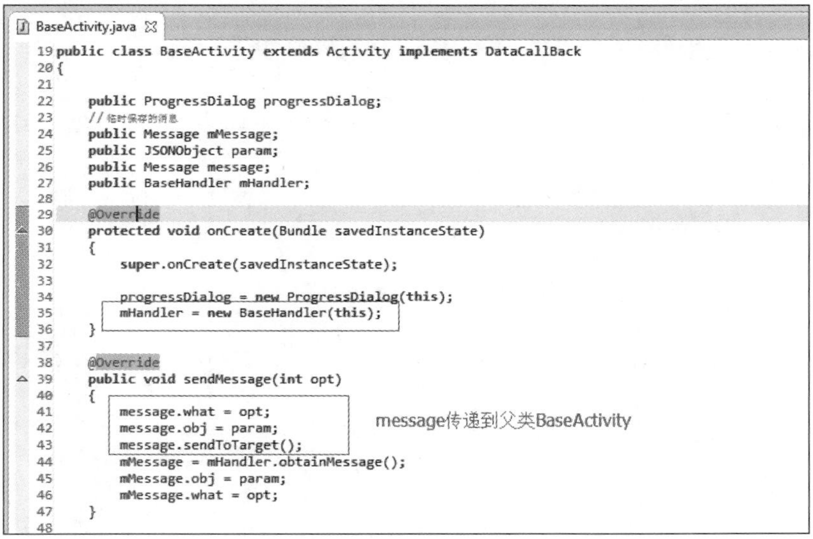

图 4.36　BaseActivity 实现 DataCallBack 接口

5）Message 对象通过 handlerMessage 传递到 MeThread

BaseHandler 重写 handleMessage 获取 Message 对象。通过判断 Message 对象的属性，开启请求服务端或者返回 Activity 数据。在启动线程 Thread 时，把 Message 对象的 what 属性赋值给 Thread 对象的 mAction 属性。同时，把 Message 对象的 obj 属性赋值给 Thread 对象的 mData 属性。代码如下：

```
package com.me.demo.handler;

import org.json.JSONObject;

import android.os.Handler;
import android.os.Message;
```

```java
import com.me.demo.dataface.DataCallBack;
import com.me.demo.thread.MeThread;

public class BaseHandler extends Handler
{
    protected DataCallBack callBack;
    private MeThread thread =null;

    public BaseHandler(DataCallBack callBack)
    {
        this.callBack =callBack;
    }

    @Override
    public void handleMessage(Message msg)
    {
        if (!Thread.currentThread().isInterrupted())
        {
            //正数代表请求服务端
            if (msg.what >0)
            {
                thread =new MeThread(this);
                thread.mAction =msg.what;
                thread.mData = (JSONObject) msg.obj;
                thread.start();
            } else
            //非正数代表服务端返回数据
            {
                callBack.getData(-msg.what, msg);
            }
        }
    }
}
```

6) Mehread 获得 Handler 后，得到 Json 数据

代码如下：

```java
public MeThread(BaseHandler handler)
{
    this.mHandler =handler;
    this.mData =new JSONObject();
}
```

7) 重写 Thread 的 Run 方法

Thread 对象调用 start 方法，启动线程，运行 Thread 中的 Run 方法。将 mAction 值

与 OPT 常量进行对比，找到对应的注册方法 userRegister。

```java
@Override
public void run()
{
    switch (mAction)
    {
        case OPT.USER_REGISTER:
            userRegister();
            break;
        default:
            break;
    }
    super.run();
}
```

8）调用 userRegister 方法

Tools 的静态方法 json2MultipartEntity 将 mData 中的 JSONObject 对象进行解析并封装到 MultipartEntity 对象中，对于解析过程，后面会详细介绍。

把 MultipartEntity 对象封装到 HttpPost 中，传递到服务端，服务端检查用户是否被注册。若未被注册，则把注册信息保存到数据库，并返回注册成功后的用户信息。返回的数据是 String 类型，实际是 JSON 数据的 String 格式，再将 String 对象转换为 JSONObject 对象。将 BeanFactory 转换为 UserBean 文件，并传递给 Handler。通过 Handler 的 handleMessage 方法调用 DataCallBack 的 getData 方法。代码如下：

```java
private void userRegister()
{
    try
    {
        String mianurl = API_URL + "user.php";
        HttpPost httpPost = new HttpPost(mianurl);
        MultipartEntity mulentity = Tools.json2MultipartEntity(mData);
        httpPost.setEntity(mulentity);
        String builder = Tools.sendMsg(httpPost);
        if (builder == null)
        {
            sendException(TIPS_GET_DATA_FAIL);
            return;
        }
        JSONObject object = new JSONObject(builder);
        if (object != null)
        {
            int flag = object.getInt("flag");
            if (flag == 1)
```

```java
            {
                JSONObject user_result =(JSONObject) object.get("userInfo");
                UserBean userBean = (UserBean) BeanFactory.json2Bean(user_result, UserBean.class);
                sendHandlerMsg(-OPT.USER_REGISTER, userBean);
            } else
            {
                String message;
                if (object.has("msg"))
                {
                    message =object.getString("msg");
                } else
                {
                    message ="";
                }
                sendException(message);
            }
        }
    } catch (JSONException e1)
    {
        e1.printStackTrace();
    }
}
```

9）用户根据服务端返回的数据情况变更界面显示

getData 方法与 sendMessage 方法类似，继承于 BaseActivity，重写此方法。当 DataCallBack 调用此方法时，显示到当前 Activity 并做出对应变化。通过 ProgressDialog 的 dismiss 方法，用户会看到相应的变化。这里只是注册功能，仅用 Toast 类在页面提示"注册成功"。代码如下：

```java
@Override
public Object getData(int opt, Message msg)
{
    if (progressDialog !=null && progressDialog.isShowing())
    {
        progressDialog.dismiss();
    }
    switch (opt)
    {
        case OPT.USER_REGISTER:
            UserBean userBean =(UserBean) msg.obj;
            Toast.makeText(this, "注册成功", Toast.LENGTH_LONG).show();
            finish();
            break;
```

```
        }
        return super.getData(opt, msg);
    }
```

4.4.2 几个关键的类

在界面类 Activity 和 Fragment 中，将部分公共的类属性、公共的方法等提出到自定义的 BaseAcitivty 和 BaseFragment 中，让代码结构更明晰。BaseActivity.java 类图如图 4.37 所示。

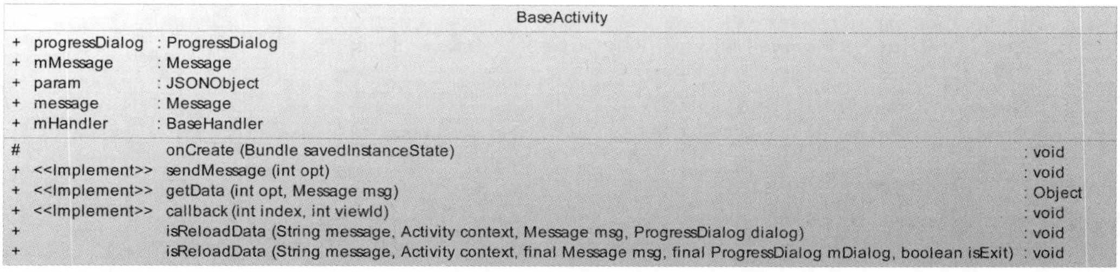

BaseActivity
+ progressDialog : ProgressDialog
+ mMessage : Message
+ param : JSONObject
+ message : Message
+ mHandler : BaseHandler
# onCreate (Bundle savedInstanceState) : void
+ <<Implement>> sendMessage (int opt) : void
+ <<Implement>> getData (int opt, Message msg) : Object
+ <<Implement>> callback (int index, int viewId) : void
+ isReloadData (String message, Activity context, Message msg, ProgressDialog dialog) : void
+ isReloadData (String message, Activity context, final Message msg, final ProgressDialog mDialog, boolean isExit) : void

图 4.37 BaseActivity.java 类图

自定义的 Dialog 对象 progressDialog 用于展现当前正在请求数据。mHandler 是自定义的 Handler，传输 Message 对象 message。message 的 obj 属性被赋值 param，what 属性被赋值 opt。Message 对象 mMessage 是为重新请求保存的临时数据。代码如下：

```
public ProgressDialog progressDialog;
//临时保存的消息，请求失败时重新请求
public Message mMessage;
public JSONObject param;
public Message message;
public BaseHandler mHandler;
@Override
protected void onCreate(Bundle savedInstanceState)
{
    super.onCreate(savedInstanceState);
    progressDialog = new ProgressDialog(this);
    mHandler = new BaseHandler(this);
}
```

BaseHandler 类图如图 4.38 所示。DataCallBack 接口图如图 4.39 所示。

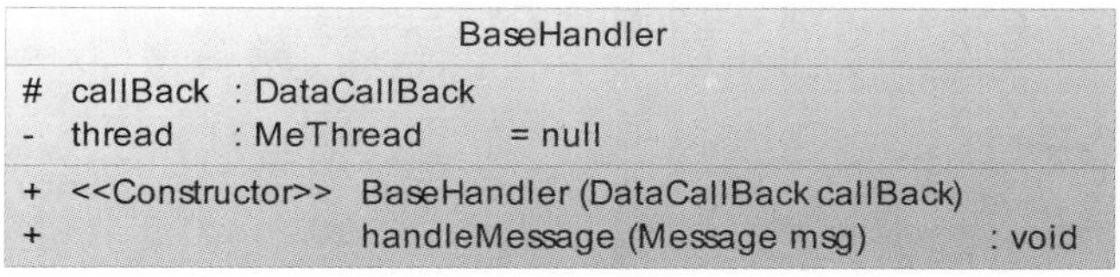

BaseHandler
callBack : DataCallBack
- thread : MeThread = null
+ <<Constructor>> BaseHandler (DataCallBack callBack)
+ handleMessage (Message msg) : void

图 4.38 BaseHandler 类图

```java
@Override
public void sendMessage(int opt)
{
    message.what = opt;
    message.obj = param;
    message.sendToTarget();
    mMessage = mHandler.obtainMessage();
    mMessage.obj = param;
    mMessage.what = opt;
}
```

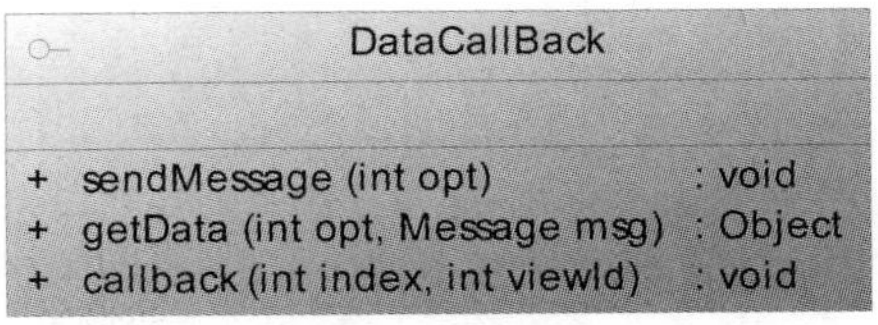

图 4.39 DataCallBack 接口图

在 BaseHandler.java 中启动一个 thread 与服务端进行交互，代码如下：

```java
package com.me.demo.handler;

import org.json.JSONObject;
import android.os.Handler;
import android.os.Message;
import com.me.demo.dataface.DataCallBack;
import com.me.demo.thread.MeThread;

public class BaseHandler extends Handler
{
    protected DataCallBack callBack;
    private MeThread thread = null;

    public BaseHandler(DataCallBack callBack)
    {
        this.callBack = callBack;
    }
    @Override
    public void handleMessage(Message msg)
    {
        if (!Thread.currentThread().isInterrupted())
        {
            if (msg.what > 0)
            {
                thread = new MeThread(this);
                thread.mAction = msg.what;
                thread.mData = (JSONObject) msg.obj;
                thread.start();
            } else
            {
                callBack.getData(-msg.what, msg);
            }
```

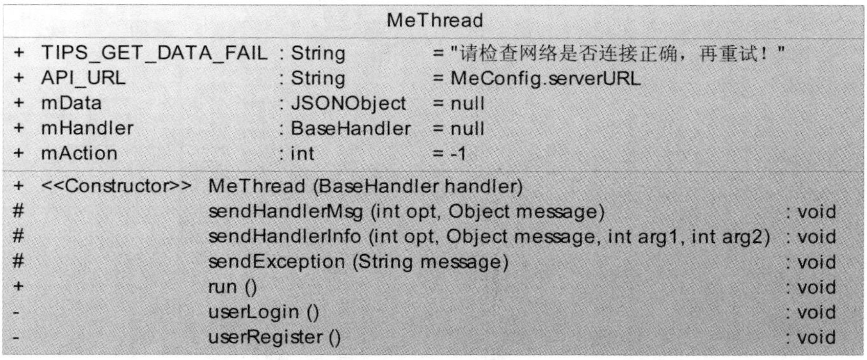

            }
        }
}
```

MeThread.java 类图如图 4.40 所示。

| MeThread | | |
|---|---|---|
| + TIPS_GET_DATA_FAIL : String | = "请检查网络是否连接正确,再重试!" | |
| + API_URL : String | = MeConfig.serverURL | |
| + mData : JSONObject | = null | |
| + mHandler : BaseHandler | = null | |
| + mAction : int | = -1 | |
| + <<Constructor>> MeThread (BaseHandler handler) | | |
| # sendHandlerMsg (int opt, Object message) | : void | |
| # sendHandlerInfo (int opt, Object message, int arg1, int arg2) | : void | |
| # sendException (String message) | : void | |
| + run () | : void | |
| - userLogin () | : void | |
| - userRegister () | : void | |

图 4.40　MeThread.java 类图

MeThread 启动后,根据参数请求不同的服务端接口。代码如下:

```java
public MeThread(BaseHandler handler)
{
    this.mHandler =handler;
    this.mData =new JSONObject();
}
protected void sendHandlerMsg(int opt, Object message)
{
    sendHandlerInfo(opt, message, 0, 0);
}
protected void sendHandlerInfo(int opt, Object message, int arg1, int arg2)
{
    Message msg =mHandler.obtainMessage();
    msg.what =opt;
    msg.obj =message;
    msg.arg1 =arg1;
    msg.arg2 =arg2;
    msg.sendToTarget();
}
protected void sendException(String message)
{
    if (message ==null)
    {
        message ="error";
    }
    Message msg =mHandler.obtainMessage();
```
```

```java
            msg.obj =message;
            msg.what =-100;
            msg.sendToTarget();
    }
    @Override
    public void run()
    {
        switch (mAction)
        {
            case OPT.USER_REGISTER:
                userRegister();
                break;
            default:
                break;
        }
        super.run();
    }
    /**
     * 注册
     */
    private void userRegister()
    {
        try
        {
            String mianurl =API_URL +"user.php";
            HttpPost httpPost =new HttpPost(mianurl);
            MultipartEntity mulentity =Tools.json2MultipartEntity(mData);
            httpPost.setEntity(mulentity);
            String builder =Tools.sendMsg(httpPost);
            if (builder ==null)
            {
                sendException(TIPS_GET_DATA_FAIL);
                return;
            }
            JSONObject object =new JSONObject(builder);
            if (object !=null)
            {
                int flag =object.getInt("flag");
                if (flag ==1)
                {
                    JSONObject user_result = (JSONObject) object.get("userInfo");
                    UserBean userBean = (UserBean) BeanFactory.json2Bean(user_result, UserBean.class);
```

```
                sendHandlerMsg(-OPT.USER_REGISTER, userBean);
            } else
            {
                String message;
                if (object.has("msg"))
                {
                    message = object.getString("msg");
                } else
                {
                    message = "";
                }
                sendException(message);
            }
        }
    } catch (JSONException e1)
    {
        e1.printStackTrace();
    }
}
```

UserBean.java 类图如图 4.41 所示。BeanFactory.java 类图如图 4.42 所示。

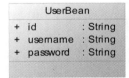

图 4.41　UserBean.java 类图　　　　图 4.42　BeanFactory.java 类图

通过简单工厂模式与反射实现 JSON 与 Bean 文件的互相转换。设计模式与反射会在知识扩展中介绍。代码如下：

```
public class BeanFactory
{
    public static final int USER = 1;
    public static final int PRODUCT = 2;

    @SuppressWarnings("rawtypes")
    public static Object json2Bean(JSONObject json, Class cls)
    {
        Object object = null;
        try
        {
            object = cls.newInstance();
        } catch (InstantiationException e1)
        {
```

```java
            e1.printStackTrace();
        } catch (IllegalAccessException e1)
        {
            e1.printStackTrace();
        }
        if (json != null)
        {
            Field[] fields = cls.getFields();
            String fname;
            try
            {
                for (Field field : fields)
                {
                    fname = field.getName();
                    String value = json.has(fname) ? json.getString(fname) : "";
                    field.set(object, value);
                }
            } catch (JSONException e)
            {
                e.printStackTrace();
            } catch (IllegalArgumentException e)
            {
                e.printStackTrace();
            } catch (IllegalAccessException e)
            {
                e.printStackTrace();
            }
        }
        return object;
    }
}
```

4.4.3 AndroidManifest.xml

1. 版本号控制

android：versionCode 是版本代码，android：versionName 是显示的版本名。

```xml
<manifest xmlns:android="http://schemas.android.com/apk/res/android"
    package="com.me.demo"
    android:versionCode="1"
    android:versionName="1.0" >
</manifest>
```

2. Android 系统支持兼容性设置

android：minSdkVersion 表示兼容的最低版本。例如 14 表示 Android 系统要 4.0 以上才可以安装这个应用。

android：targetSdkVersion 设置编译的版本。

```
<uses-sdk
    android:minSdkVersion="14"
    android:targetSdkVersion="19" />
```

3. 权限控制

我们这个应用要访问网络、磁盘、地理位置等，都需要通过系统的权限受理。访问网络前，必须在 AndroidManifest.xml 文件中配置如下权限。

```
<uses-permission android:name="android.permission.INTERNET" />
```

其他权限见附录 A。

4. 注册与配置

在 application 内主要针对 Android 系统 4 大组件，进行注册与配置。

```
<application
    android:allowBackup="true"
    android:icon="@drawable/ic_launcher"
    android:label="@string/app_name"
    android:theme="@style/AppTheme" >
    <activity
        android:name=".activity.StartActivity"
        android:label="@string/app_name" >
        <intent-filter>
            <action android:name="android.intent.action.MAIN" />
            <category android:name="android.intent.category.LAUNCHER" />
        </intent-filter>
    </activity>
    <activity android:name=".activity.MainActivity" >
    </activity>
    <activity android:name=".activity.RegisterActivity" >
    </activity>
    <activity android:name=".activity.LoginActivity" >
    </activity>
</application>
```

4.5 用户注册功能的调试

（1）将 meServer 服务端程序复制到 Tomcat 的 webapps 目录下，双击 bin 目录中的 startup.bat 启动 Tomcat。

（2）确保手机 WiFi 与计算机在同一个局域网络中，通过 CMD 命令查询 IP 地址，如图 4.43 所示。

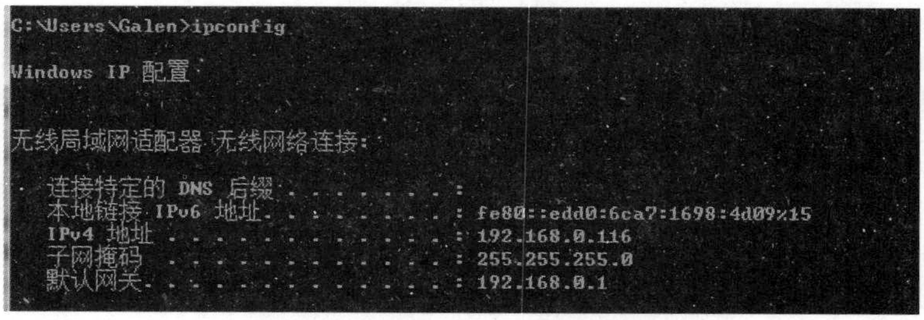

图 4.43 计算机 IP 地址

（3）修改 MeConfig.java 的 serverURL。代码如下：

```
package com.me.demo.util;
public class MeConfig
{
    public static final String serverURL ="http://192.168.0.116:8080/meServer/";
}
```

（4）运行 MeDemo 程序，进入注册页面。

（5）在用户界面输入用户名 test、密码 123456，再重复输入密码 123456，单击"注册"按钮。

（6）提示注册成功，打开 Navicat 查看 me_server 中的 me_user 表，如图 4.44 所示。

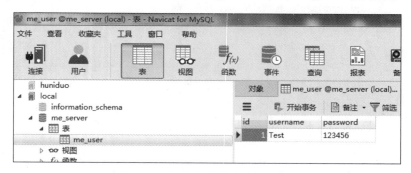

图 4.44 注册成功后写到数据库中

4.6 知识点回顾与技能扩展

4.6.1 知识点回顾

本章主要知识点如下：

（1）熟练掌握 Android 项目的目录结构。

（2）熟练掌握 xml 布局文件的创建与五大布局类型。

（3）熟练掌握 Activity 的创建，以及 Activity 的生命周期。

（4）熟练掌握 dimen 资源的使用。

（5）熟练掌握 drawable 资源文件的使用。

（6）熟练掌握 AndroidManifest.xml 的结构和使用。

（7）精通 Activity 的使用。

4.6.2 技能扩展

1. TextView 属性

TextView 属性如表 4.4 所示。

表 4.4 TextView 属性

属 性 名	描 述
android：ems="N"	设置宽度为 N 个字符
android：maxems="N"	设置宽度最长为 N 个字符，与 ems 同时使用时覆盖 ems 选项
android：minems="N"	设置 TextView 的宽度最短为 N 个字符，与 ems 同时使用时覆盖 ems 选项
android：maxLength="N"	限制输入字符数，如设置为 5
android：lines="N"	设置文本的行数
android：maxLines	设置文本的最大显示行数
android：minLines	设置文本的最小行数
android：lineSpacingExtra	设置行间距
android：lineSpacingMultiplier	设置行间距的倍数，如"1.2"
android：password	以小点"."显示文本
android：phoneNumber	设置电话号码的输入方式
android：singleLine	设置单行显示
android：textAppearance	设置文字外观
android：textColor	设置文本颜色

续表

属 性 名	描 述
android：textColorHighlight	被选中文字的底色
android：textColorHint	设置提示信息文字的颜色
android：textColorLink	文字链接的颜色
android：textScaleX	设置文字之间的间隔
android：textSize	设置文字大小
android：textStyle	设置字型(bold(粗体)：0，italic(斜体)：1，bolditalic(又粗又斜)：2)，可以设置一个或多个
android：typeface	设置文本字体
android：height	设置文本区域的高度
android：maxHeight	设置文本区域的最大高度
android：minHeight	设置文本区域的最小高度
android：width	设置文本区域的宽度
android：maxWidth	设置文本区域的最大宽度
android：minWidth	设置文本区域的最小宽度

2. EditText 属性

由于 EditText 继承自 TextView，所以 EditText 具有 TextView 属性的特点，下面主要介绍一些 EditText 特有的输入法属性的特点。EditText 属性如表 4.5 所示。

表 4.5　EditText 属性

属 性 名	描 述
android：inputType="none"	输入类型无限制
android：inputType="text"	文本输入
android：inputType="textCapCharacters"	输入普通字符
android：inputType="textCapWords"	单词首字母大写
android：inputType="textCapSentences"	仅第一个字母大写
android：inputType="textAutoCorrect"	自动检测拼写
android：inputType="textAutoComplete"	前两个自动完成
android：inputType="textMultiLine"	多行输入
android：inputType="textImeMultiLine"	输入法多行(不一定支持)
android：inputType="textNoSuggestions"	不提示
android：inputType="textUri"	URI 格式

续表

属 性 名	描 述
android：inputType＝"textEmailAddress"	电子邮件地址格式
android：inputType＝"textEmailSubject"	邮件主题格式
android：inputType＝"textShortMessage"	短消息格式
android：inputType＝"textLongMessage"	长消息格式
android：inputType＝"textPersonName"	人名格式
android：inputType＝"textPostalAddress"	邮政格式
android：inputType＝"textPassword"	密码格式
android：inputType＝"textVisiblePassword"	密码可见格式
android：inputType＝"textWebEditText"	作为网页表单的文本格式
android：inputType＝"textFilter"	文本筛选格式
android：inputType＝"textPhonetic"	拼音输入格式
android：inputType＝"number"	数字格式
android：inputType＝"numberSigned"	有符号数字格式
android：inputType＝"numberDecimal"	可以带小数点的浮点格式
android：inputType＝"phone"	拨号键盘
android：inputType＝"datetime"	日期时间
android：inputType＝"date"	日期键盘
android：inputType＝"time"	时间键盘

3. Button 属性

Button 属性如表 4.6 所示。

表 4.6 Button 属性

属 性 名	描 述
android：autoLink	控制链接网址和电子邮件地址等是否自动发现并转换为可单击的链接
android：bufferType	确定最低类型 getText()并返回
android：capitalize	如果设置为 true,将自动纠正输入值的拼写；如果设置为 false 或者不设置（默认）则不纠正
android：cursorVisible	使得光标（默认）可见或不可见
android：drawableBottom	在 text 的下方输出一个 drawable,可以是图片、样式、颜色等
android：drawableLeft	在 text 的左边输出一个 drawable,可以是图片、样式、颜色等

续表

属 性 名	描 述
android：drawablePadding	设置 text 与 drawable 的间距，与 drawableLeft、drawableRight、drawableTop、drawableBottom 一起使用
android：drawableRight	在 text 的右边输出一个 drawable，可以是图片、样式、颜色等
android：drawableTop	在 text 的正上方输出一个 drawable，可以是图片、样式、颜色等
android：editorExtras	设置文本额外的输入数据
android：ellipsize	设置当文字过长时，该控件该如何显示。start：省略号显示在开头，end：省略号显示在结尾，middle：省略号显示在中间，marquee：以跑马灯的方式显示（动画横向移动）
android：ems	设置 TextView 的宽度为 N 个字符
android：freezesText	设置保存文本的内容以及光标的位置
android：gravity	设置文本位置，如设置成 center，文本将居中显示
android：hint	Text 为空时显示的文字提示信息
android：imeActionId	为 EditorInfo 提供一个值，actionId 时使用一个输入连接到文本视图方法
android：imeActionLabel	为 EditorInfo 提供一个值，actionLabel 时使用一个输入连接到文本视图方法
android：imeOptions	启用一个输入法与一个编辑器来提高与应用程序的集成
android：inputMethod	为文本指定输入法，需要完全限定名（完整的包名）
android：inputType	文本字段的数据类型
android：lineSpacingExtra	设置行间距
android：lineSpacingMultiplier	设置行间距的倍数
android：lines	设置文本的行数
android：linksClickable	设置是否点击链接
android：marqueeRepeatLimit	在 ellipsize 指定 marquee 的情况下，设置重复滚动的次数，当设置为 marquee_forever 时表示无限次
android：maxEms	设置 TextView 的宽度为最长为 N 个字符的宽度与 ems 同时使用时覆盖 ems 选项
android：maxHeight	设置文本区域的最大高度
android：maxLines	设置文本的最大显示行数，与 width 或者 layout_width 结合使用，超出部分会自动换行，超出行数将不显示
android：maxWidth	设置文本区域的最大宽度
android：minEms	设置 TextView 的宽度最短为 N 个字符
android：minHeight	设置文本区域的最小高度
android：minLines	设置文本的最小行数，与 lines 类似

续表

属 性 名	描 述
android：minWidth	设置文本区域的最小宽度
android：phoneNumber	设置电话号码的输入方式
android：privateImeOptions	设置输入法选项
android：scrollHorizontally	设置文本超出 TextView 宽度的情况下，是否出现横拉条
android：selectAllOnFocus	如果文本是可选择的，则获取焦点而不是将光标移动到文本的开始位置或末尾位置
android：shadowColor	指定文本阴影的颜色，需要与 shadowRadius 一起使用
android：shadowDx	设置阴影横向坐标开始位置
android：shadowDy	设置阴影纵向坐标开始位置
android：shadowRadius	设置阴影的半径。设置为 0.1 就变成字体的颜色了，一般设置为 3.0 的效果比较好
android：textAppearance	设置文字外观
android：textColor	设置文本颜色
android：textColorHighlight	被选中文字的底色，默认为蓝色
android：textColorHint	提示文本的颜色
android：textColorLink	链接文本的颜色
android：textIsSelectable	表明的内容不可编辑的文本可以选择
android：textScaleX	设置文本的水平扩展因素
android：textSize	设置文本大小
android：textStyle	设置字体风格（粗体、斜体、***bolditalic***）文本
android：setTypeface	设置（正常、无衬线、等宽字体）文本
android：layout_width	设置父布局允许 view 所占的宽度

4．布局的灵活性与布局原则

考虑到效果图的功能，交互性要提供更好的体验，还要考虑到页面的性能，因此布局需要灵活性，但考虑到性能，则必须遵守一些布局原则。

布局的原则：布局的优化主要是深度和广度，深度的表现主要在于布局的嵌套使用，广度的表现主要是包含过多的视图。

（1）避免不必要的嵌套。尽量不要把一个布局放置在其他布局里面。

（2）避免使用太多视图。在一个布局中每增加一个新的视图，都会在 inflate 操作中耗时和消耗资源。任何时候都不要在一个布局中包含超过 80 个视图，否则，消耗在 inflate 操作上的时间会很多。

（3）避免深度嵌套。布局可以任意嵌套，这很容易创建复杂和深度嵌套的布局层次。

如果没有硬件限制，最好将嵌套限制在 10 层以下。

（4）尽量多使用相对布局 RelativeLayout，不要使用绝对布局 AbsoluteLayout。由于 Android 的碎片化程度很高，市面上存在的屏幕尺寸各式各样，使用 RelativeLayout 能使构建的布局适应性更强，构建出来的 UI 布局对多屏幕的适配效果较好。通过指定 UI 控件间的相对位置，可使在不同屏幕上布局的表现基本保持一致。当然，也不是所有情况下都得使用相对布局，要根据具体情况选择和其他布局方式的搭配来实现最优布局。

（5）将可复用的组件抽取出来并通过＜include /＞标签使用。如将注册顶部组成部分抽取出来，在注册的 layout 中如下调用：

```
<include
    android:layout_width="match_parent"
    android:layout_height="@dimen/action_bar_default_height"
    layout="@layout/top_layout_register_activity" />
```

（6）使用＜ViewStub/＞标签来加载一些不常用的布局。ViewStub 调用 inflate() 方法或设置 visible 之前，它是不占用布局空间和系统资源的。它的使用场景可以是在需要加载并显示一些不常用的 View 时，例如一些网络异常的提示信息等。

（7）使用＜merge/＞标签减少布局的嵌套层次。该标签的主要使用场景主要分为两种情况。

第一种情况是当 xml 文件的根布局是 FrameLayout 时，可以用 merge 作为根节点，这是因为 Activity 的内容布局中，默认就用了一个 FrameLayout 作为 xml 布局根节点的父节点。main.xml 的根节点是一个 RelativeLayout，其父节点就是一个 FrameLayout。如果在 main.xml 里面使用 FrameLayout 作为根节点，就可以使用 merge 合并成一个 FrameLayout，这样就降低了布局嵌套层次。

第二种情况是当用 include 标签导入一个共用布局时，如果父布局和子布局根节点为同一类型，就可以使用 merge 将子节点布局的内容合并到父布局中，这样就可以减少一级嵌套层次。首先看看不使用 merge 的情况。新建一个布局文件 commonnaviright.xml 用来构建一个在导航栏右边的按钮布局。

```
<include
        android:layout_toLeftOf="@id/image_top_layout_right"
        android:layout_toRightOf="@id/image_top_layout_left"
        layout="@layout/merge_layout_top_text_title" />
```

merge_layout_top_text_title.xml 内容如下：

```
<?xml version="1.0" encoding="utf-8"?>
<merge xmlns:android="http://schemas.android.com/apk/res/android"
    xmlns:tools="http://schemas.android.com/tools"
    android:layout_width="match_parent"
    android:layout_height="match_parent" >
    <TextView
```

```
        android:id="@+id/text_top_title"
        android:layout_width="match_parent"
        android:layout_height="wrap_content"
        android:layout_centerInParent="true"
        android:gravity="center_horizontal"
        android:text="@string/register"
        android:textSize="@dimen/text_size_18" />
</merge>
```

4.7 练　　习

实现某手机论坛的用户注册。

第 5 章 用户登录

5.1 用户登录总体设计

登录时序如图 5.1 所示。

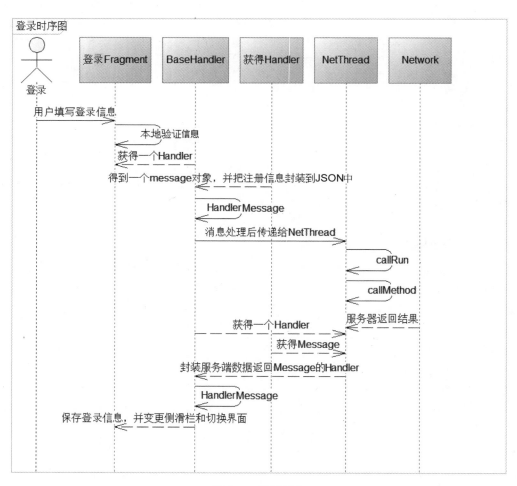

图 5.1 登录时序

登录流程如图 5.2 所示。

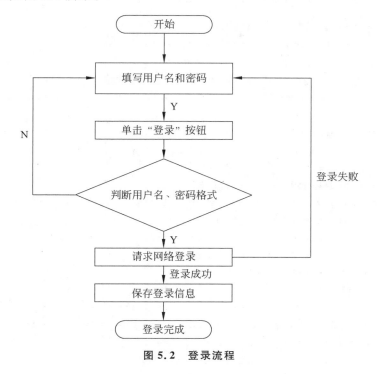

图 5.2　登录流程

5.2　用户登录的实现

5.2.1　登录的具体实现

1．登录界面效果及实现

登录界面如图 5.3 所示。

图 5.3　登录界面

顶部布局：top_layout_login_activity.xml。代码如下：

```xml
<?xml version="1.0" encoding="utf-8"?>
<RelativeLayout xmlns:android="http://schemas.android.com/apk/res/android"
    android:layout_width="match_parent"
    android:layout_height="@dimen/action_bar_default_height"
    android:background="@drawable/top_layout_supplier_action_bar" >
    <TextView
        android:id="@+id/text_top_layout_right"
        android:layout_width="@dimen/action_bar_default_height"
        android:layout_height="match_parent"
        android:layout_alignParentRight="true"
        android:gravity="center_vertical|center_horizontal"
        android:text="@string/register_"
        android:textColor="@android:color/holo_blue_dark"
        android:textSize="@dimen/text_size_20"
        android:textStyle="bold" />
    <ImageView
        android:id="@+id/image_top_layout_left"
        android:layout_width="@dimen/action_bar_default_height"
        android:layout_height="match_parent"
        android:layout_alignParentLeft="true"
        android:contentDescription="@string/app_name"
        android:padding="8dip"
        android:src="@drawable/icon_back" />
    <RelativeLayout
        android:layout_width="match_parent"
        android:layout_height="match_parent"
        android:layout_toLeftOf="@id/text_top_layout_right"
        android:layout_toRightOf="@id/image_top_layout_left"
        android:orientation="vertical" >
        <TextView
            android:id="@+id/text_top_title"
            android:layout_width="match_parent"
            android:layout_height="wrap_content"
            android:layout_centerInParent="true"
            android:gravity="center_horizontal"
            android:text="@string/login"
            android:textSize="@dimen/text_size_18" />
    </RelativeLayout>
</RelativeLayout>
```

页面布局：activity_login.xml。代码如下：

```xml
<?xml version="1.0" encoding="utf-8"?>
```

```xml
<LinearLayout xmlns:android="http://schemas.android.com/apk/res/android"
    xmlns:tools="http://schemas.android.com/tools"
    android:layout_width="match_parent"
    android:layout_height="match_parent"
    android:background="@color/white"
    android:orientation="vertical"
    tools:context="com.me.demo.fragment.LoginFragment" >
    <include
        android:layout_width="match_parent"
        android:layout_height="@dimen/action_bar_default_height"
        layout="@layout/top_layout_login_activity" />
    <LinearLayout
        android:layout_width="match_parent"
        android:layout_height="wrap_content"
        android:layout_margin="10dp"
        android:padding="5dp" >
        <TextView
            android:layout_width="wrap_content"
            android:layout_height="wrap_content"
            android:text="@string/user_name"
            android:textSize="@dimen/text_size_18" />
        <EditText
            android:layout_width="match_parent"
            android:layout_height="wrap_content"
            android:background="@null"
            android:hint="@string/user_name_tips" />
    </LinearLayout>
    <View
        android:layout_width="match_parent"
        android:layout_height="1dp"
        android:layout_marginBottom="5dp"
        android:layout_marginLeft="10dp"
        android:layout_marginRight="10dp"
        android:layout_marginTop="5dp"
        android:background="@android:color/darker_gray" />
    <LinearLayout
        android:layout_width="match_parent"
        android:layout_height="wrap_content"
        android:layout_margin="10dp"
        android:orientation="horizontal"
        android:padding="5dp" >
        <TextView
            android:layout_width="wrap_content"
            android:layout_height="wrap_content"
```

```xml
            android:text="@string/password"
            android:textSize="@dimen/text_size_18" />
        <EditText
            android:layout_width="match_parent"
            android:layout_height="wrap_content"
            android:background="@null"
            android:hint="@string/password_tips" >
        </EditText>
    </LinearLayout>
    <Button
        android:layout_width="match_parent"
        android:layout_height="wrap_content"
        android:layout_margin="10dp"
        android:background="@drawable/bkg_button"
        android:padding="10dp"
        android:text="@string/login"
        android:textColor="@color/white"
        android:textSize="@dimen/text_size_20" />
</LinearLayout>
```

2. 登录流程控制

1) LoginActivity 类图

LoginActivity.java 类图如图 5.4 所示。

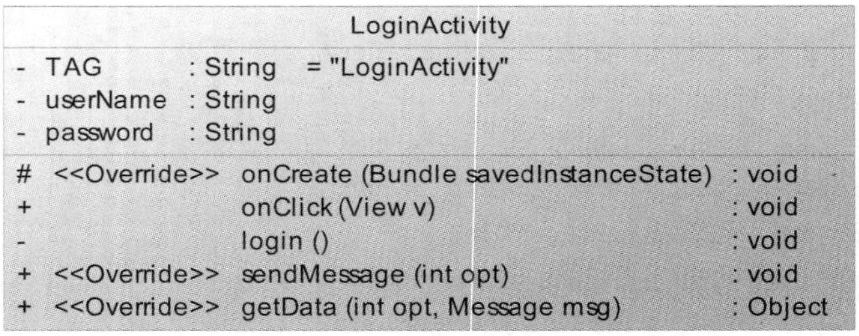

LoginActivity			
- TAG	: String	= "LoginActivity"	
- userName	: String		
- password	: String		
# <<Override>>	onCreate (Bundle savedInstanceState)		: void
+	onClick (View v)		: void
+	login ()		: void
+ <<Override>>	sendMessage (int opt)		: void
+ <<Override>>	getData (int opt, Message msg)		: Object

图 5.4　LoginActivity.java 类图

2) 组件监听事件

LoginActivity 监听用户单击登录按钮的操作。

LoginActivity 实现 android.view.View.OnClickListener 接口，这是单击事件监听的接口。

public class LoginActivity **extends** BaseActivity **implements** OnClickListener

重写 run 方法，v.getId()将获得组件的 id 值。代码如下：

@Override
public void onClick(View v)

```java
{
    switch (v.getId())
    {
        case R.id.button_login:
            login();
            break;
        default:
            break;
    }
}
```

登录按钮设置单击监听事件，代码如下：

```java
findViewById(R.id.button_login).setOnClickListener(this);
```

3）获取登录相关的数据信息

单击登录按钮后，执行login()方法。代码如下：

```java
private void login()
{
    //检查用户名格式
    EditText userNameEdit = (EditText) findViewById(R.id.edit_username_login);
    userName =userNameEdit.getText() +"";
    if (!userName.trim().matches("^[a-zA-Z0-9_][a-zA-Z0-9_]{3,10}$"))
    {
        WidgetTools.setTVError(userNameEdit, this.getResources().getString(R.string.toast_user_name_tips), this);
        return;
    }
    //检查密码格式
    EditText passwordEdit = (EditText) findViewById(R.id.edit_password_login);
    password =passwordEdit.getText() +"";
    if (!password.trim().matches("^[\\@A-Za-z0-9\\!\\#\\$\\%\\^\\&\\*\\.\\~]{5,22}$"))
    {
        WidgetTools.setTVError (passwordEdit, getResources().getString(R.string.toast_user_password_tips), this);
        return;
    }
    sendMessage(OPT.USER_LOGIN);
}
```

4）封装登录信息

封装登录信息到JSON，为了提高用户的体验，我们在请求接口时，progressDialog 提示用户正在请求服务器。代码如下：

```java
@Override
public void sendMessage(int opt)
{
    progressDialog.show();
    message = mHandler.obtainMessage();
    param = new JSONObject();
    try
    {
        switch (opt)
        {
            case OPT.USER_LOGIN:
                param.put("act", "login");
                param.put("username", userName);
                param.put("password", password);
                break;
        }
    } catch (JSONException e)
    {
        e.printStackTrace();
    }
    super.sendMessage(opt);
}
```

5）传递登录信息

传递登录信息与注册流程相同。经过 BaseActivity 传递信息到 BaseHandler，再传递到 Mehread，调用 run()方法，执行 userLogin 方法。代码如下：

```java
private void userLogin()
{
    try
    {
        String mianurl = API_URL + "user.php";
        HttpPost httpPost = new HttpPost(mianurl);
        MultipartEntity mulentity = Tools.json2MultipartEntity(mData);
        httpPost.setEntity(mulentity);
        String builder = Tools.sendMsg(httpPost);
        if (builder == null)
        {
            sendException(TIPS_GET_DATA_FAIL);
            return;
        }
        JSONObject object = new JSONObject(builder);
        if (object != null)
```

```
            {
                int flag = object.getInt("flag");
                if (flag == 1)
                {
                    JSONObject user_result = (JSONObject) object.get("userInfo");
                    UserBean userBean = (UserBean) BeanFactory.json2Bean(user_result, UserBean.class);
                    sendHandlerMsg(-OPT.USER_LOGIN, userBean);
                } else
                {
                    String message;
                    if (object.has("msg"))
                    {
                        message = object.getString("msg");
                    } else
                    {
                        message = "";
                    }
                    sendException(message);
                }
            }
        } catch (JSONException e1)
        {
            sendException(TIPS_GET_DATA_FAIL);
            e1.printStackTrace();
        }
    }
```

6）注册 LoginActivity

在 Manifest.xml 中注册 LoginActivity。代码如下：

```
<activity android:name=".activity.LoginActivity" ></activity>
```

5.2.2 客户端与服务器的交互

将数据封装到 JSON 对象中。代码如下：

```
param.put("act", "login");
param.put("username", userName);
param.put("password", password);
```

服务端返回 JSON String 解析放到 javabean 文件 UserBean.java 中。代码如下：

```
package com.me.demo.bean;

public class UserBean
```

```
{
    public String id;
    public String username;
    public String password;
}
```

5.2.3 后台服务接口文档

接口地址：http://localhost:8080/meServer/user.php。

调用方式：POST。

登录接口参数如表 5.1 所示。

表 5.1 登录接口参数

请求参数	必选	类型及范围	说 明
act	Y	String	login
username	Y	String	用户名
password	Y	String	密码

返回方式：JSON。

调用示例：http://localhost:8080/meServer/user.php?act=login&username=123&password=321。

登录接口返回 JSON 数据如表 5.2 所示。

表 5.2 登录接口返回 JSON 数据

返回值字段	字段类型	字段说明
flag	Int	1 成功,0 失败
msg	String	消息提示信息
userInfo	JSONObject	用户信息 Json 数据
id	String	用户 id
password	String	用户密码
username	String	用户名

5.3 用户登录的调试

(1) 准备工作请参照 4.7 节。

(2) 若登录成功,则 Toast 提示登录成功,否则提示登录失败。

5.4 支持用户使用第三方账号登录

5.4.1 什么是第三方登录

第三方登录就是利用用户在第三方平台上已有的账号来快速完成自己应用的登录或者注册的功能。这里的第三方平台，一般是已经有大量用户的平台，如国内的新浪微博、QQ，国外的Facebook、Twitter等。第三方登录不是一个具体的接口，而是一种思想或者一套步骤。

要实现第三方登录，首先需要选择一个第三方平台。新浪微博和QQ空间都是好的选择。这些平台拥有大量的用户，而且还开放了API，供我们调用接入。但是同样开放API，微信却不是一个好选择，这是因为微信的API只支持分享，不支持授权验证或者获取用户资料。因此要实现第三方登录，所选择的平台至少需要具备以下特性：

(1) 开放API。

(2) 可获取用户资料或至少可进行授权验证。

ShareSDK已经支持了超过20种这样的平台，完全足够选择使用。

5.4.2 第三方账号登录方式

选择好平台后，现在思考的问题是：你的应用是否具备独立账户系统？

这个问题是第三方登录时接口选择的重要标准。如果选择"是"，则意味着应用只是需要第三方平台的用户，而不是其账户验证功能——也就是"要数据，不要功能"。如果选择"否"，则表示实际上是"要功能，不要数据（用户）"。对于ShareSDK来说，前者的入口方法是showUser(null)，而后者的入口方法是authorize()。它主要有两种接入方式。

(1) 要功能，不要数据。

(2) 要数据，不要功能。

5.4.3 使用第三方账号登录

1. 要功能，不要数据

如果你的应用不具备用户系统，而且也不打算维护这个系统，那么可以依照下面的步骤来做：

(1) 用户触发第三方登录事件。

(2) 调用platform.getDb().getUserId()，请求用户在此平台上的ID。

(3) 如果用户ID存在，则认为该用户是合法用户，允许进入系统；否则调用authorize()。

(4) authorize()方法引导用户在授权页面输入账号、密码，目标平台将验证此用户。

(5) 如果 onComplete()方法被回调,则表示授权成功,引导用户进入系统;否则提示错误,调用 removeAccount()方法,删除可能的授权缓存数据。

2. 要数据,不要功能

如果你的应用拥有用户系统,就是说该应用自己就有注册和登录功能,使用第三方登录只是为了拥有更多用户,那么可以依照下面的步骤来做:

(1) 用户触发第三方登录事件。

(2) showUser(null)请求授权用户的资料(这个过程中可能涉及授权操作)。

(3) 如果 onComplete()方法被回调,则将其参数 Hashmap 代入该应用的 Login 流程;否则提示错误,调用 removeAccount()方法,删除可能的授权缓存数据。

(4) Login 时客户端发送用户资料中的用户 ID 给服务端。

(5) 服务端判定用户是否已注册。若已注册,则引导用户进入系统,否则返回特定错误码。

(6) 客户端收到"未注册用户"错误码以后,代入用户资料到你的应用的 Register 流程。

(7) Register 时在用户资料中挑选该应用的注册所需字段,并提交服务端注册。

(8) 服务端完成用户注册。若成功,则反馈给客户端引导用户进入系统;否则提示错误,调用 removeAccount()方法,删除可能的授权缓存数据。

5.5 知识点回顾与技能扩展

5.5.1 知识点回顾

本章主要知识点如下:

(1) LinearLayout 的布局样式与基本属性。

(2) RelativeLayout 的布局样式与基本属性。

(3) Activity 的生命周期。

(4) 与后台服务端的交互。

5.5.2 技能扩展

1. Java 语言的反射机制

1) 概述

Java 语言的反射机制是指:在运行状态中,对于任意一个类,都能够知道这个类的所有属性和方法;对于任意一个对象,都能够调用它的任意一个方法和属性。这种动态获取信息以及动态调用对象的方法称为 Java 语言的反射机制。

2) 功能

Java 语言的反射机制主要提供了以下功能:在运行时判断任意一个对象所属的类;

在运行时构造任意一个类的对象；在运行时判断任意一个类所具有的成员变量和方法；在运行时调用任意一个对象的方法；生成动态代理。

有时我们说某个语言具有很强的动态性，有时我们会区分动态和静态的不同技术与做法，经常用到动态绑定（dynamic binding）、动态链接（dynamic linking）、动态加载（dynamic loading）等。"动态"一词其实没有绝对而普遍适用的严格定义，有时甚至像面向对象当初被导入编程领域一样，一人一把号，各吹各的调。

通常，开发者社群说到动态语言，大致认同的一个定义是："程序运行时，允许改变程序结构或变量类型，这种语言称为动态语言"。从这个观点看，Perl、Python、Ruby 是动态语言，C++、Java、C#不是动态语言。

虽然在这样的定义与分类下 Java 不是动态语言，但它却有着一个非常突出的动态相关机制：Reflection。其意思是"反射、映像、倒影"，用在 Java 上指的是：可以于运行时加载、探知、使用编译期间完全未知的 Class。换句话说，Java 程序可以加载一个运行时才得知名称的 Class，获悉其完整构造（但不包括 Method 定义），并生成其对象实体，或对其 Field 赋值，或唤起其 Method。这种"看透 Class"的能力（the ability of the program to examine itself）被称为 Introspection（内省、内观、反省）。Reflection 和 Introspection 是常被并提的两个术语。

Java 如何能够具有上述动态特性呢？这是一个深远的话题，本文对此只简单介绍一些概念，主要还是介绍 Reflection APIs，也就是让读者知道如何探索 Class 的结构，如何对某个"运行时才获知名称的 Class"生成一份实体，为其 Field 设值，调用其 Method。本文将谈到 java.lang.Class，以及 java.lang.reflect 中的 Method、Field、Constructor 等 Class。

3) Class 类

Java 程序在运行时，Java 运行时系统一直对所有的对象进行所谓的运行时类型标识。这项信息记录了每个对象所属的类。虚拟机通常使用运行时类型信息选择正确的方法去执行，用来保存这些类型信息的类是 Class 类。

也就是说，ClassLoader 找到了需要调用的类时（Java 为了调控内存的调用消耗，类的加载都在需要时再进行，这样做很有效），就会加载它，然后根据.class 文件内记载的类信息来产生一个与该类相联系的独一无二的 Class 对象。该 Class 对象记载了该类的字段、方法等信息。以后 JVM 要产生该类的实例，就根据内存中存在的该 Class 类所记载的信息（Class 对象应该和其他类一样会在堆内存内产生、消亡）来进行。

Java 中的 Class 类对象是可以人工自然（即是开放的）得到的（虽然无法像其他类一样运用构造器来得到其实例，因为 Class 对象都是 JVM 产生的；不过，由客户产生也是无意义的），而且更重要的是，基于这个基础，Java 实现了反射机制。

Class 类中存在以下几个重要的方法：

（1）getName()：Class 对象描述了一个特定类的特定属性，而这个方法就是返回 String 形式的该类的简要描述。由于历史原因，对数组的 Class 对象调用该方法会产生奇怪的结果。

（2）newInstance()：该方法可以根据某个 Class 对象产生其对应类的实例。需要强

调的是,它调用的是此类的默认构造方法。例如:

```
Object x = new Object();
Object y = x.getClass().newInstance();
```

(3) getClassLoader():返回该 Class 对象对应的类的类加载器。

(4) getComponentType():该方法针对数组对象的 Class 对象,可以得到该数组的组成元素所对应对象的 Class 对象。例如:

```
int[] ints = new int[]{1,2,3};
Class class1 = ints.getClass();
Class class2 = class1.getComponentType();
```

这里得到的 class2 对象所对应的就是 int 这个基本类型的 Class 对象。

(5) getSuperClass():返回某子类所对应的直接父类所对应的 Class 对象。

(6) isArray():判定此 Class 对象所对应的是否是一个数组对象。

2. JSON 数据介绍

1) 什么是 JSON

JSON(JavaScript Object Notation)是一种轻量级的数据交换格式。因为解析 XML 比较复杂,而且需要编写大段代码,所以客户端和服务器的数据交换往往通过 JSON 来进行。尤其是对于 Web 开发来说,JSON 数据格式在客户端可以直接通过 JavaScript 进行解析。

JSON 有两种数据结构,其中一种是以(key/value)对的形式存在的无序的 jsonObject 对象。一个对象以"{"(左花括号)开始,以"}"(右花括号)结束。每个"名称"后跟一个":"(冒号);"名称/值"对之间使用","(逗号)分隔。

- JSONObject 格式:{string:value}或者{string,value}。比如{{"name":"安卓"}},就是一个简单的 JSON 对象。key 值必须是 string 类型,而对于 value,则可以是 string、number、object、array 等数据类型。
- JSONArray 格式:{string:[value,value,...]}

有关 JSON 格式可参考 JSON 的官网:http://www.json.org/json-zh.html。

2) JSON 数据在项目中的使用

Android 的 JSON 解析部分都在包 org.json 下,主要有以下几个类:

- JSONObject:可以看作一个 JSON 对象。
- JSONStringer:JSON 文本构建类。
- JSONArray:可以看作 JSON 的数组。
- JSONTokener:JSON 解析类。
- JSONException:JSON 中用到的异常。

注册用户时,要对数据进行封装。代码如下:

```
@Override
public void sendMessage(int opt) {
```

```java
            progressDialog.show();
            message = mHandler.obtainMessage();
            //最外层是{},创建一个对象
    param = new JSONObject();
    try {
        switch (opt) {
        case OPT.USER_REGISTER:
            //如果第一个键的值是数组,则需要创建数组对象
            //JSONArray array = new JSONArray();
            //param.put("array", array);
            param.put("act", "register");
            param.put("username", userName);
            param.put("password", password);
            break;
        }
    } catch (JSONException e) {
        e.printStackTrace();
    }
    super.sendMessage(opt);
}
```

服务端获取到我们传递的 JSON 数据后,完成相关操作,返回给客户端的数据也是 JSON 数据格式。我们通过解析 JSON 数据格式,再用反射将数据保存到 JavaBean 的对象中。

反射解析 JavaBean 成员变量,解析 JSON 数据。代码如下:

```java
/**
 *
 * @param json JSONObject 对象
 * @param cls JavaBean 的类
 * @return 该 JavaBean 的对象
 */
@SuppressWarnings("rawtypes")
public static Object json2Bean(JSONObject json, Class cls)
{
    Object object = null;
    try
    {
        //创建 JavaBean 文件的对象
        object = cls.newInstance();
    } catch (InstantiationException e1)
    {
        e1.printStackTrace();
    } catch (IllegalAccessException e1)
```

```java
        {
            e1.printStackTrace();
        }
        if (json !=null)
        {
            //通过反射,获取JavaBean文件的成员变量,并封装到数组中
            Field[] fields =cls.getFields();
            String fname;
            try
            {
                //遍历数组,查询JSON对象,若成功则赋值给JavaBean的成员变量
                for (Field field : fields)
                {
                    fname =field.getName();
                    String value =json.has(fname) ?json.getString(fname) : "";
                    field.set(object, value);
                }
            } catch (JSONException e)
            {
                e.printStackTrace();
            } catch (IllegalArgumentException e)
            {
                e.printStackTrace();
            } catch (IllegalAccessException e)
            {
                e.printStackTrace();
            }
        }
        //返回该JavaBean的对象
        return object;
}
```

服务端返回的数据为 JSON 格式的 String。代码如下：

```java
//服务端返回JSON格式的String builder
JSONObject object =new JSONObject(builder);
if (object !=null)
{
    int flag =object.getInt("flag");
    if (flag ==1)
    {
        JSONObject user_result = (JSONObject) object.get("userInfo");
        UserBean userBean = (UserBean) BeanFactory.json2Bean(user_result,
UserBean.class);
        sendHandlerMsg(-OPT.USER_LOGIN, userBean);
```

```
    } else
    {
        String message;
        if (object.has("msg"))
        {
            message =object.getString("msg");
        } else
        {
            message ="";
        }
        sendException(message);
    }
}
```

3. SQLite 介绍

SQLite 是一个非常流行的嵌入式数据库，它支持 SQL 语言，并且只利用很少的内存就会有很好的性能。此外它还是开源的，任何人都可以使用它。许多开源项目（如 Mozilla、PHP、Python）都使用了 SQLite。

SQLite 基本上符合 SQL-92 标准，和其他主要的 SQL 数据库没什么区别。它的优点就是高效。Android 运行时环境包含了完整的 SQLite。

SQLite 和其他数据库最大的不同就是对数据类型的支持。创建一个表时，可以在 CREATE TABLE 语句中指定某列的数据类型，但可以把任何数据类型放入任何列中。当某个值插入数据库时，SQLite 将检查其类型。如果该类型与关联的列不匹配，则 SQLite 会尝试将该值转换成该列的类型。如果不能转换，则该值将作为其本身具有的类型存储。比如可以把一个字符串(String)放入 INTEGER 列。SQLite 称其为"弱类型" (manifest typing)。

此外，SQLite 不支持某些标准的 SQL 功能，特别是外键约束（foreign key constrain）、嵌套 transcaction、right outer join 和 full outer join，还有 alter table 功能。

总之，SQLite 是一个完整的 SQL 系统，拥有完整的触发器等。

4. 关于设计模式

1）概述

设计模式(Design Pattern)是一套被反复使用、多数人知晓、经过分类编目的代码设计经验的总结。使用设计模式是为了可重用代码，让代码更容易被他人理解，以便保证代码的可靠性。毫无疑问，设计模式于己、于他人、于系统是多赢的。设计模式可使代码编制真正工程化。它是软件工程的基石，如同大厦的结构一样。设计模式是一门高深的学问。如果要细说起来，恐怕一本书也难以完全写清楚。简单来讲，设计模式提供了很多软件工程问题的解决方案。一般来说，常用的 Android 设计模式有以下 8 种：单例、工厂、观察者、代理、命令、适配器、合成、访问者。

由于设计模式博大精深,下面仅介绍几种初学者入门的设计模式。

2) 单例模式

单例模式又分为两种:懒汉模式和饿汉模式。单例模式就是永远保持一个对象。懒汉模式在运行的时候 获取对象比较慢,但是加载类的时候比较快。饿汉模式是在运行的时候获取对象较快,加载类的时候较慢。饿汉模式是线程安全的,在类创建的同时就已经创建好一个静态的对象供系统使用,以后不再改变;懒汉模式如果在创建实例对象时不加上 synchronized,则会导致对对象的访问不是线程安全的。

懒汉模式:只有在自身需要的时候才会行动,从来不知道及早做好准备。它在需要对象的时候,才判断是否已有对象。如果没有就立即创建一个对象,然后返回;如果已有对象就不再创建,立即返回。懒汉模式只在外部对象第一次请求实例的时候才去创建。

示例:

```java
public class LazySingleton
{
    private static LazySingleton m_instance =null;
    /**
     * 私有的构造子,外界无法直接实例化
     */
    private LazySingleton() { }
    /**
     * 静态工厂方法,返回唯一实例
     */
    synchronized public static LazySingleton getInstance()
    {
        if (m_instance ==null)
        {
            m_instance =new LazySingleton();
        }
        return m_instance;
    }
}
```

饿汉模式:加载这个类的时候,立即创建。

示例:

```java
public class EagerSingleton
{
    private static final EagerSingleton m_instance =new EagerSingleton();
    /**
     * 私有的构造函数
     */
```

```java
    private EagerSingleton() { }
    /**
     * 静态工厂方法
     */
    public static EagerSingleton getInstance()
    {
        return m_instance;
    }
}
```

3) 简单工厂模式

简单工厂模式(Simple Factory Pattern)属于类的创新型模式,又叫静态工厂方法模式(Static Factory Method Pattern)。它通过专门定义一个类来负责创建其他类的实例,被创建的实例通常都具有共同的父类。定义一个用于创建对象的接口,让子类决定实例化哪个类。工厂模式使一个类的实例化延迟到其子类。

在 Android 中,创建 Bitmap 对象的时候经常使用静态工厂方法,例如通过资源 id 获取 Bitmap 对象。

BitmapFactory 的工厂方法具体实现代码如下:

```java
/**
 * Synonym for {@link #decodeResource(Resources, int, android.graphics.BitmapFactory.Options)}
 * will null Options.
 *
 * @param res The resources object containing the image data
 * @param id The resource id of the image data
 * @return The decoded bitmap, or null if the image could not be decode.
 */
public static Bitmap decodeResource(Resources res, int id) {
    return decodeResource(res, id, null);
}


/**
 * Synonym for opening the given resource and calling
 * {@link #decodeResourceStream}.
 *
 * @param res    The resources object containing the image data
 * @param id The resource id of the image data
 * @param opts null-ok; Options that control downsampling and whether the
 *             image should be completely decoded, or just is size returned.
 * @return The decoded bitmap, or null if the image data could not be
 *         decoded, or, if opts is non-null, if opts requested only the
 *         size be returned (in opts.outWidth and opts.outHeight)
 */
```

```java
public static Bitmap decodeResource(Resources res, int id, Options opts) {
    Bitmap bm = null;
    InputStream is = null;

    try {
        final TypedValue value = new TypedValue();
        is = res.openRawResource(id, value);

        bm = decodeResourceStream(res, value, is, null, opts);
    } catch (Exception e) {
        /*  do nothing.
            If the exception happened on open, bm will be null.
            If it happened on close, bm is still valid.
        */
    } finally {
        try {
            if (is != null) is.close();
        } catch (IOException e) {
            //Ignore
        }
    }

    if (bm == null && opts != null && opts.inBitmap != null) {
        throw new IllegalArgumentException("Problem decoding into existing bitmap");
    }

    return bm;
}
```

其中，decodeResource(Resources res, int id)函数调用 decodeResource(Resources res, int id, Options opts)。在 decodeResource 函数中，把传递进来的资源 id 解析成 InputStream，然后调用 decodeResourceStream(res, value, is, null, opts)方法。

decodeResourceStream 的实现代码如下：

```java
/**
 * Decode a new Bitmap from an InputStream. This InputStream was obtained from
 * resources, which we pass to be able to scale the bitmap accordingly.
 */
public static Bitmap decodeResourceStream(Resources res, TypedValue value,
        InputStream is, Rect pad, Options opts) {

    if (opts == null) {
        opts = new Options();
```

```java
        }

        if (opts.inDensity ==0 && value !=null) {
            final int density =value.density;
            if (density ==TypedValue.DENSITY_DEFAULT) {
                opts.inDensity =DisplayMetrics.DENSITY_DEFAULT;
            } else if (density !=TypedValue.DENSITY_NONE) {
                opts.inDensity =density;
            }
        }

        if (opts.inTargetDensity ==0 && res !=null) {
            opts.inTargetDensity =res.getDisplayMetrics().densityDpi;
        }

        return decodeStream(is, pad, opts);
    }

    /**
    * Decode an input stream into a bitmap. If the input stream is null, or
    * cannot be used to decode a bitmap, the function returns null.
    * The stream's position will be where ever it was after the encoded data
    * was read.
    *
    * @param is The input stream that holds the raw data to be decoded into a
    *           bitmap.
    * @param outPadding If not null, return the padding rect for the bitmap if
    *                   it exists, otherwise set padding to [-1,-1,-1,-1]. If
    *                   no bitmap is returned (null) then padding is
    *                   unchanged.
    * @param opts null-ok; Options that control downsampling and whether the
    *             image should be completely decoded, or just is size returned.
    * @return The decoded bitmap, or null if the image data could not be
    *         decoded, or, if opts is non-null, if opts requested only the
    *         size be returned (in opts.outWidth and opts.outHeight)
    */
    public static Bitmap decodeStream (InputStream is, Rect outPadding, Options opts) {
        //we don't throw in this case, thus allowing the caller to only check
        //the cache, and not force the image to be decoded.
        if (is ==null) {
            return null;
        }
```

```java
        //we need mark/reset to work properly

        if (!is.markSupported()) {
            is = new BufferedInputStream(is, 16 * 1024);
        }

        //so we can call reset() if a given codec gives up after reading up to
        //this many bytes. FIXME: need to find out from the codecs what this
        //value should be.
        is.mark(1024);

        Bitmap  bm;

        if (is instanceof AssetManager.AssetInputStream) {
              bm = nativeDecodeAsset (((AssetManager.AssetInputStream) is).getAssetInt(),
                 outPadding, opts);
        } else {
            //pass some temp storage down to the native code. 1024 is made up,
            //but should be large enough to avoid too many small calls back
            //into is.read(...) This number is not related to the value passed
            //to mark(...) above.
            byte [] tempStorage = null;
            if (opts != null) tempStorage = opts.inTempStorage;
            if (tempStorage == null) tempStorage = new byte[16 * 1024];
            bm = nativeDecodeStream(is, tempStorage, outPadding, opts);
        }
        if (bm == null && opts != null && opts.inBitmap != null) {
             throw new IllegalArgumentException ("Problem decoding into existing bitmap");
        }

        return finishDecode(bm, outPadding, opts);
}

private static Bitmap finishDecode(Bitmap bm, Rect outPadding, Options opts)
      {
    if (bm == null || opts == null) {
        return bm;
    }

    final int density = opts.inDensity;
    if (density == 0) {
        return bm;
```

```java
        }
        bm.setDensity(density);
        final int targetDensity = opts.inTargetDensity;
        if (targetDensity == 0 || density == targetDensity || density == opts.inScreenDensity) {
            return bm;
        }

        byte[] np = bm.getNinePatchChunk();
        final boolean isNinePatch = np != null && NinePatch.isNinePatchChunk(np);
        if (opts.inScaled || isNinePatch) {
            float scale = targetDensity / (float)density;
            //TODO: This is very inefficient and should be done in native by Skia
            final Bitmap oldBitmap = bm;
            bm = Bitmap.createScaledBitmap(oldBitmap, (int) (bm.getWidth() * scale +0.5f),
                    (int) (bm.getHeight() * scale +0.5f), true);
            oldBitmap.recycle();

            if (isNinePatch) {
                np = nativeScaleNinePatch(np, scale, outPadding);
                bm.setNinePatchChunk(np);
            }
            bm.setDensity(targetDensity);
        }

        return bm;
    }
```

decodeResourceStream 初始化一些配置和像素信息后,调用 decodeStream(is,pad,opts),最终调用 nativeDecodeAsset 或者 nativeDecodeStream 来构建 Bitmap 对象,这两个都是 native 方法(Android 中使用 Skia 库来解析图像)。调用 finishDecode 函数来设置像素、配置信息、缩放等参数,最终返回 Bitmap 对象。

BitmapFactory 中的 decodeFile、decodeByteArray 工厂方法都是类似的过程,BitmapFactory 通过不同的工厂方法和传递不同的参数调用不同的图像解析函数来构造 Bitmap 对象。

5.6 练　　习

(1) 实现第 3 章已注册用户的登录。
(2) 实现用户使用 QQ 账号登录。

第6章 向用户展示内容

6.1 基本内容展示总体设计

内容展示时序如图 6.1 所示。

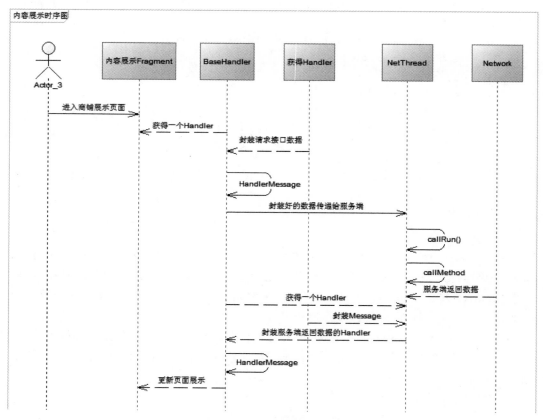

图 6.1 内容展示时序

内容展示流程如图 6.2 所示。

第 6 章 向用户展示内容

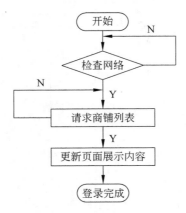

图 6.2 内容展示流程

6.2 数据库准备

6.2.1 数据库商户

数据库商户结构如图 6.3 所示。

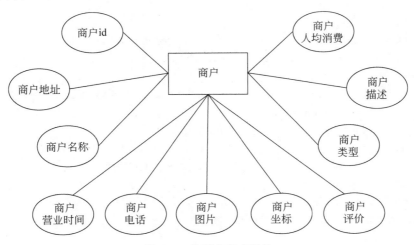

图 6.3 数据库商户结构

6.2.2 数据库商户表

数据库商户表如表 6.1 所示。

表 6.1 数据库商户表

名称	说明	数据类型	主键/外键/非空
shop_id	商户 id	int	P
shop_address	商户地址	varchar	
shop_name	商户名称	varchar	
shop_time	商户营业时间	timestamp	
shop_phone	商户电话	varchar	
shop_image	商户图片	varchar	
shop_longitude	商户经度	varchar	
shop_latitude	商户纬度	varchar	
shop_type	商户类型	varchar	
shop_desc	商户描述	varchar	
shop_spend	商户人均消费	varchar	

新建一张表 me_shop.sql，代码如下。

```sql
DROP TABLE IF EXISTS 'me_shop';
CREATE TABLE 'me_shop' (
    'shop_id' int(11) unsigned NOT NULL AUTO_INCREMENT,
    'shop_address' varchar(120) NOT NULL,
    'shop_name' varchar(80) NOT NULL,
    'shop_time' timestamp NOT NULL DEFAULT CURRENT_TIMESTAMP,
    'shop_phone' varchar(20) NOT NULL,
    'shop_image' varchar(120) NOT NULL,
    'shop_longitude' varchar(16) NOT NULL COMMENT '经度',
    'shop_latitude' varchar(16) NOT NULL COMMENT '纬度',
    'shop_type' varchar(2) NOT NULL,
    'shop_desc' varchar(255) NOT NULL,
    'shop_spend' varchar(12) NOT NULL,
    PRIMARY KEY ('shop_id')
) ENGINE=InnoDB AUTO_INCREMENT=51 DEFAULT CHARSET=utf8;
```

手动录入部分数据到 me_shop 表中。调试请参照 4.7 节。

6.2.3 后台服务端接口文档

接口地址：http://localhost:8080/meServer/shop.php。

调用方式：POST。

请求商铺列表接口参数如表 6.2 所示。

表 6.2 请求商铺列表接口参数

请 求 参 数	必　选	类型及范围	说　　明
act	Y	String	getShops
pageSize	Y	String	每页显示数量
page	Y	String	从 1 开始,逐渐增加

返回方式:JSON。

调用示例:http://localhost:8080/meServer/shop.php?act=getShops&pageSize=20&page=1。

请求商铺接口列表返回参数如表 6.3 所示。

表 6.3 请求商铺接口列表返回参数

返回值字段	字 段 类 型	字 段 说 明
flag	Int	1:成功,0:失败
msg	String	消息提示信息
vendorList	JSONOArray	店铺列表数据
distance	String	距离店铺距离(只有传入当前经、纬度才会显示距离,定位将在后面介绍)
shopAddress	String	店铺地址
shopCreateTime	String	店铺创建时间
shopDesc	String	店铺描述
shopId	String	店铺 id
shopImage	String	店铺图片
shopLatitude	String	店铺经度
shopLongitude	String	店铺纬度
shopName	String	店铺名称
shopPhone	String	店铺电话
shopSpend	String	店铺人均消费
shopTime	String	店铺营业时间
shopType	String	店铺类型

6.3 内容展示知识点详解

6.3.1 Fragment 介绍

Fragment 所在包：android.app.Fragment，android.support.v4.app.Fragment（引用 android-support-v4.jar 这个包时）。

Fragment 是 Android honeycomb 3.0 新增的组件，它和 Activity 十分相似。在一个 Activity 中，Fragment 用来描述一些行为或一部分用户界面。可以合并多个 Fragment，在一个单独的 Activity 中，建立多个 UI 面板，同时在多个 Activity 中重用 Fragment。Fragment 作为一个 Activity 中的模块，有自己的生命周期，接收自己的输入事件，从运行中的 Activity 可以添加或移除 Fragment。

Fragment 必须嵌入在一个 Activity 中，Fragment 的生命周期受 Activity 的影响。

下面介绍 Fragment 的几个重要方法。

1. onAttach(Activity)

当 Fragment 与 Activity 发生关联时调用该方法。

2. onCreateView(LayoutInflater inflater, ViewGroup, Bundle)

创建该 Fragment 的视图，加载布局文件一般在此方法中调用。

```
@Override
public View onCreateView(LayoutInflater inflater, ViewGroup container, Bundle savedInstanceState)
{
    rootView = inflater.inflate(R.layout.fragment_main, container, false);
    rootView.findViewById(R.id.image_top_layout_left).setOnClickListener(this);
    vfRecomm = (ViewFlipper) rootView.findViewById(R.id.hots_vf_recomm);
    textFooterMsg = (TextView) rootView.findViewById(R.id.textFooterMsg);
    vendorList.clear();
    vfRecomm.removeAllViews();
    recommPage = 1;
    page = 1;
    isLastLoad = false;
    sendMessage(OPT.GET_SHOPS);
    return rootView;
}
```

3. onActivityCreated(Bundle)

当 Activity 的 onCreate 方法返回时调用。

4. onDestoryView()

与 onCreateView 相对应,当该 Fragment 的视图被移除时调用。

5. onDetach()

与 onAttach 相对应,当 Fragment 与 Activity 的关联被取消时调用。

Fragment 的生命周期如图 6.4 所示。Fragment 与 Activity 生命周期的关系如图 6.5 所示。

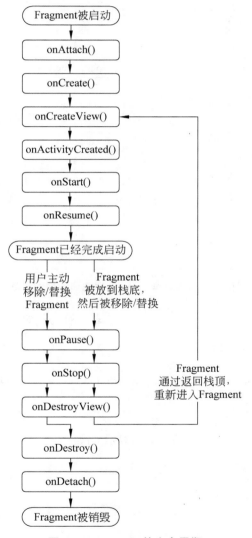

图 6.4 Fragment 的生命周期

如何创建 Fragment 并具体实现?我们来看看 MainFragment 是如何实现的。MainFragment 是从 MainActivity.java 类启动的,即 MainFragment"寄生"在 MainActivity 中。

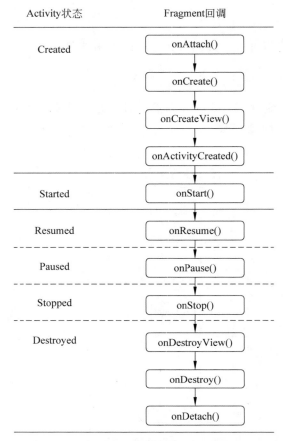

图 6.5　Fragment 与 Activity 生命周期的关系

在 onCreate(Bundle savedInstanceState)方法中的代码如下：

```
if (mainFragment ==null)
    {
        mainFragment =MainFragment.newInstance(MainFragment.STATE_PICS);
    }
transact(mainFragment);
```

transact(BaseFragment mFragment)方法代码如下：

```
private void transact(BaseFragment mFragment)
{
    FragmentTransaction transaction =getSupportFragmentManager().beginTransaction();
    transaction.replace(R.id.content_frame, mFragment);
    transaction.addToBackStack(null);
    transaction.commit();
}
```

6.3.2 FragmentManage 介绍

FragmentManage 所在包：android. app. FragmentManager, android. support. v4. app. FragmentManager(引用 android-support-v4. jar 这个包时)。

FragmentManager 能够管理 Activity 中的 Fragment。

FragmentManager 的几个重要方法如下：

1. beginTransaction()

获取一个 FragmentTransaction 对象。

2. getFragmentManager()

通过调用 Activity 的 getFragmentManager()取得其实例。

3. findFragmentById() 或 findFragmentByTag()

获取 Activity 中已存在的 Fragment。

4. popBackStack()

从 Activity 的后退栈中弹出 Fragment。

5. addOnBackStackChangedListener()

用方法 addOnBackStackChangedListener()注册一个侦听器,以监视后退栈的变化。

6.3.3 FragmentTransaction 介绍

FragmentTransaction 所在包：android. app. FragmentTransaction, android. support. v4. app. FragmentManager(引用 android-support-v4. jar 这个包时)。

FragmentTransaction(事务)对 Fragment 进行添加、移除、替换,以及执行其他动作。

FragmentTransaction 的几个重要方法如下。

1. add（int containerViewId，Fragment fragment，String tag）

add 将一个 Fragment 添加到一个容器 container 里。

2. replace（int containerViewId，Fragment fragment，String tag）

replace 先移除相同 id 的所有 Fragment,然后再添加当前的 Fragment。

在大部分情况下,上述两个方法的表现基本相同。因为一般会使用一个 FrameLayout 作为容器,而每个 Fragment 被添加或者替换到这个 FrameLayout 的时候,都是显示在最上层的,所以所看到的界面都是一样的。在使用 add 的情况下,这个

FrameLayout 其实有两层，多层肯定要比一层浪费资源，所以还是推荐使用 replace。当然，有时候还是需要使用 add 的。比如要实现轮播图的效果，每个轮播图都是一个独立的 Fragment，而其容器 FrameLayout 需要添加多个 Fragment，这样就可以根据提供的逻辑进行轮播了。

3. addToBackStack()

该方法添加当前 transaction 到恢复栈中。这意味着当前 transaction 在提交之后将会被存储，在它被出栈时将会恢复所做的操作。

4. commit()

该方法安排一个 transaction 的提交操作，但不会立即执行。

6.4 内 容 展 示

6.4.1 内容展示的具体实现

1. 内容展示界面的效果

1) 首页

首页展示商铺的效果如图 6.6 所示。

fragment_main.xml 的布局组成：顶部（Layout）、中间滑动组件（ViewFlipper）、底部（TextView）显示页码。

（1）滑动组件（ViewFlipper）。

android：layout_width="299dp"，android：layout_height="397dp"固定容器 ViewFlipper 的宽、高，方便 Activity 动态布局。

android：layout_centerHorizontal="true"设置 ViewFlipper 始终居中显示。

android：flipInterval="1000"设置切换布局时的动画时间间隔，单位为毫秒。

```
<ViewFlipper
    android:id="@+id/hots_vf_recomm"
    android:layout_width="299dp"
    android:layout_height="397dp"
    android:layout_centerHorizontal="true"
    android:flipInterval="1000"
    android:persistentDrawingCache="animation" >
</ViewFlipper>
```

图 6.6 首页展示商铺的效果

(2) 底部(TextView)。

android：textAppearance＝"？android：attr/textAppearanceMedium"引用系统自带的外观。

```
<TextView
    android:id="@+id/textFooterMsg"
    android:layout_width="wrap_content"
    android:layout_height="wrap_content"
    android:layout_alignParentBottom="true"
    android:layout_centerHorizontal="true"
    android:layout_marginBottom="8dp"
    android:textAppearance="?android:attr/textAppearanceMedium"
    android:textColor="#847563"
    android:textSize="13sp" />
```

2）全部商户

全部商户的效果如图 6.7 所示。

图 6.7 全部商户的效果

fragment_shops_main.xml 布局组成：顶部(Layout)、中间(ListView)。

```
<LinearLayout xmlns:android="http://schemas.android.com/apk/res/android"
    xmlns:tools="http://schemas.android.com/tools"
    android:layout_width="match_parent"
    android:layout_height="match_parent"
    android:orientation="vertical"
    tools:context="com.me.demo.fragment.MainFragment" >
```

```
<include
    android:layout_width="match_parent"
    android:layout_height="@dimen/action_bar_default_height"
    layout="@layout/top_layout_shops_fragment" />
<ListView
    android:id="@+id/list_view_shops_main"
    android:layout_width="match_parent"
    android:layout_height="match_parent" />
</LinearLayout>
```

滑动组件(ListView)代码如下:

```
<ListView
    android:id="@+id/list_view_shops_main"
    android:layout_width="match_parent"
    android:layout_height="match_parent" />
```

2．内容展示的流程控制

1）首页

MainFragment.java 类图如图 6.8 所示。

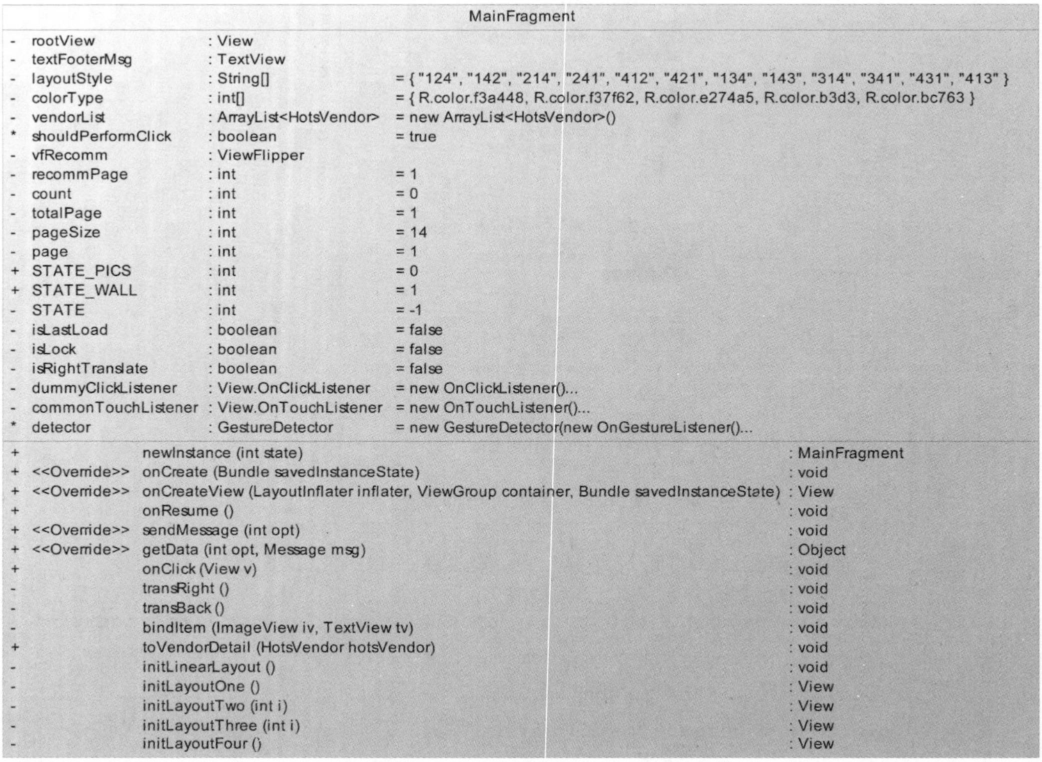

MainFragment		
- rootView	: View	
- textFooterMsg	: TextView	
- layoutStyle	: String[]	= { "124", "142", "214", "241", "412", "421", "134", "143", "314", "341", "431", "413" }
- colorType	: int[]	= { R.color.f3a448, R.color.f37f62, R.color.e274a5, R.color.b3d3, R.color.bc763 }
- vendorList	: ArrayList<HotsVendor>	= new ArrayList<HotsVendor>()
* shouldPerformClick	: boolean	= true
- vfRecomm	: ViewFlipper	
- recommPage	: int	= 1
- count	: int	= 0
- totalPage	: int	= 1
- pageSize	: int	= 14
- page	: int	= 1
+ STATE_PICS	: int	= 0
+ STATE_WALL	: int	= 1
+ STATE	: int	= -1
- isLastLoad	: boolean	= false
- isLock	: boolean	= false
- isRightTranslate	: boolean	= false
- dummyClickListener	: View.OnClickListener	= new OnClickListener()...
- commonTouchListener	: View.OnTouchListener	= new OnTouchListener()...
- detector	: GestureDetector	= new GestureDetector(new OnGestureListener())...
+ newInstance (int state)		: MainFragment
+ <<Override>> onCreate (Bundle savedInstanceState)		: void
+ <<Override>> onCreateView (LayoutInflater inflater, ViewGroup container, Bundle savedInstanceState)		: View
+ onResume ()		: void
+ <<Override>> sendMessage (int opt)		: void
+ <<Override>> getData (int opt, Message msg)		: Object
+ onClick (View v)		: void
- transRight ()		: void
- transBack ()		: void
- bindItem (ImageView iv, TextView tv)		: void
+ toVendorDetail (HotsVendor hotsVendor)		: void
- initLinearLayout ()		: void
- initLayoutOne ()		: View
- initLayoutTwo (int i)		: View
- initLayoutThree (int i)		: View
- initLayoutFour ()		: View

图 6.8 MainFragment.java 类图

进入页面,展示的内容需要动态加载到 ViewFliper 里。第一步是请求服务端获得店铺列表信息。

```java
param.put("act", "getShops");
param.put("pageSize", pageSize);
param.put("page", page);
```

将服务端数据封装到 HotsVendor 的 JavaBean 文件中。

```java
ArrayList<HotsVendor>vendorList =new ArrayList<HotsVendor>();
JSONArray jsonArray =object.getJSONArray("vendorList");
int size =jsonArray.length();
for (int i =0; i <size; i++)
{
    vendorList.add((HotsVendor) BeanFactory.json2Bean(jsonArray.getJSONObject(i), HotsVendor.class));
}
```

MainFragment 接收服务端的数据,初始化 ViewFlipper。

```java
@Override
public Object getData(int opt, Message msg)
{
    if (progressDialog !=null && progressDialog.isShowing())
    {
        progressDialog.dismiss();
    }
    switch (opt)
    {
        case OPT.GET_SHOPS:
            //若无法加载一页的数据,则停止加载数据
            ArrayList<HotsVendor>newsList = (ArrayList<HotsVendor>) msg.obj;
            if (newsList.size() <7)
            {
                Toast.makeText(getActivity(), "数据已加载完毕!", Toast.LENGTH_LONG).show();
                isLastLoad =true;
            } else
            {
                if (newsList.size() <pageSize)
                {
                    Toast.makeText(getActivity(), "数据已加载完毕!", Toast.LENGTH_LONG).show();
                    isLastLoad =true;
                } else
                {
```

```
                    page++;
                }
                vendorList.addAll(newsList);
                totalPage = vendorList.size() / 7;
            }
            if (recommPage ==1)
            {
                //第一次请求接口,初始化第二页
                initLinearLayout();
            } else if (recommPage ==totalPage)
            {
                //接口请求到最后一页,下一次翻转到第一页,不需要初始化页面
            } else
            {
                //每次请求接口后,若不是第一页也不是最后一页,则需要重新初始化页面
                recommPage++;
                initLinearLayout();
                recommPage--;
            }
            textFooterMsg.setText(recommPage + "/" +totalPage);
            isLock = false;
            if (isRightTranslate)
            {
                transRight();
                isRightTranslate = false;
            }
            break;
    }
    return super.getData(opt, msg);
}
```

对于总共页数、当前页及是否需要向下切换等属性进行定义。

```
//总共页数
private int totalPage =1;
//当前页
private int recommPage =1;
//是否需要向下切换
private boolean isRightTranslate =false;
```

显示页面的组件,重写 onCreateView 方法。用 LayoutInflater.inflater 方法装载布局。子组件需要用加载的布局去掉 findViewById 方法。View.removeAllViews()是清除所有子布局的方法。

```
@Override
public View onCreateView(LayoutInflater inflater, ViewGroup container, Bundle
```

```
    savedInstanceState)
    {
        rootView = inflater.inflate(R.layout.fragment_main, container, false);
        rootView.findViewById(R.id.image_top_layout_left).setOnClickListener
(this);
        //ViewFlipper 容器
        vfRecomm = (ViewFlipper) rootView.findViewById(R.id.hots_vf_recomm);
        //显示页面
        textFooterMsg = (TextView) rootView.findViewById(R.id.textFooterMsg);
        //商铺列表
        vendorList.clear();
        //清除内容,重置数据
        vfRecomm.removeAllViews();
        recommPage = 1;
        page = 1;
        isLastLoad = false;
        sendMessage(OPT.GET_SHOPS);
        return rootView;
    }
```

initLinearLayout()方法初始化下一页 ViewFlipper 显示内容,其中 4 个子布局文件如图 6.9 所示。

```
    //页面布局样式随机
    private String[] layoutStyle = { "124", "142", "214", "241", "412", "421", "134",
 "143", "314", "341", "431", "413" };

    private void initLinearLayout()
    {
        count = 0;
        //新建一个 LinearLayout 布局
        LinearLayout linearLayout = new LinearLayout(getActivity());
        //布局方向为垂直方向
        linearLayout.setOrientation(LinearLayout.VERTICAL);
        //设置 Layout 布局类型
        linearLayout.setBackgroundColor(getResources().getColor(R.color.fff));
        //设置内部布局居中
        linearLayout.setGravity(Gravity.CENTER);
        //设置全屏
        linearLayout. setLayoutParams (new LayoutParams (LayoutParams. MATCH_
PARENT, LayoutParams.MATCH_PARENT));
        //随机布局样式,长度为 3 行
        String style = layoutStyle[new Random().nextInt(layoutStyle.length)];
        for (int i = 0; i < style.length(); i++)
        {
```

```
            View child =null;
            switch (style.charAt(i))
            {
                case '1':
                    //以布局方式 1 布局
                    child =initLayoutOne();
                    break;
                case '2':
                    //以布局方式 2 布局
                    child =initLayoutTwo(i);
                    break;
                case '3':
                    //以布局方式 3 布局
                    child =initLayoutThree(i);
                    break;
                case '4':
                    //以布局方式 4 布局
                    child =initLayoutFour();
                    break;
            }
            linearLayout.addView(child);
        }
        //设置滑动监听事件
        linearLayout.setOnTouchListener(commonTouchListener);
        //设置单击监听事件
        linearLayout.setOnClickListener(dummyClickListener);
        //ViewFlipper 添加定义的 LinearLayout
        vfRecomm.addView(linearLayout);
    }
```

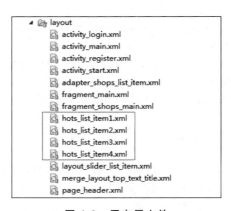

图 6.9 子布局文件

触摸屏幕监听事件，调用手势监听事件。

```java
private View.OnTouchListener commonTouchListener = new OnTouchListener()
{
    public boolean onTouch(View v, MotionEvent event)
    {
        shouldPerformClick = true;
        return detector.onTouchEvent(event);
    }
};
```

手势监听事件，主要重写 onFling 方法。

```java
GestureDetector detector = new GestureDetector(new OnGestureListener()
{
    @Override
    public boolean onSingleTapUp(MotionEvent e)
    {
        return false;
    }
    @Override
    public void onShowPress(MotionEvent e)
    {
    }
    @Override
    public boolean onScroll(MotionEvent e1, MotionEvent e2, float distanceX, float distanceY)
    {
        return false;
    }
    @Override
    public void onLongPress(MotionEvent e)
    {
    }
    private static final int SWIPE_MIN_DISTANCE = 120;
    private static final int SWIPE_THRESHOLD_VELOCITY = 200;
    @Override
    public boolean onFling(MotionEvent e1, MotionEvent e2, float velocityX, float velocityY)
    {
        boolean isConsumed = false;
        if (Math.abs(velocityY) > SWIPE_THRESHOLD_VELOCITY)
        {
            shouldPerformClick = false;
            if (STATE == STATE_PICS && !isLock)
            {
                if (e1.getY() - e2.getY() > SWIPE_MIN_DISTANCE)
```

```java
            {
                //若子布局数量等于当前页数,则初始化一个
                int childCount =vfRecomm.getChildCount();
                //ViewFlipper装载最后一页,但不是数据的最后一页,重新初始化下
                //  一页
                if (recommPage !=totalPage && recommPage ==childCount)
                {
                    recommPage++;
                    initLinearLayout();
                    recommPage--;
                    transRight();
                    isConsumed =true;
                //是请求数据的最后一页,但服务端还能请求数据,请求服务端下一页
                //  数据
                } else if (recommPage ==totalPage && !isLastLoad)
                {
                    //数据拿到后,执行翻页
                    isRightTranslate =true;
                    sendMessage(OPT.GET_SHOPS);
                    shouldPerformClick =!isConsumed;
                    return isConsumed;
                } else
                {
                    //服务端数据没有新数据,ViewFlipper也装载完毕,则转到第一页
                    transRight();
                    isConsumed =true;
                }
            } else if (e1.getY() -e2.getY() <-SWIPE_MIN_DISTANCE)
            {
                transBack();
                isConsumed =true;
            }
        }
    }
    shouldPerformClick =!isConsumed;
    return isConsumed;
}
@Override
public boolean onDown(MotionEvent e)
{
    return false;
}
});
```

正向、逆向旋转,setInAnimation 设置 ViewFlipper 动画效果,加载动画 xml 文档,如

图 6.10 所示。

```
//正向旋转
private void transRight()
{
    MainFragment.this.vfRecomm.setInAnimation(AnimationUtils.loadAnimation
(MainFragment.this.getActivity(), R.anim.push_top_in));
    MainFragment.this.vfRecomm.setOutAnimation(AnimationUtils.loadAnimation
(MainFragment.this.getActivity(), R.anim.push_top_out));
    MainFragment.this.vfRecomm.showNext();
    if (++recommPage >totalPage)
    {
        recommPage =1;
    }
    textFooterMsg.setText(recommPage +"/" +totalPage);
}
//逆向旋转
private void transBack()
{
    MainFragment.this.vfRecomm.setInAnimation(AnimationUtils.loadAnimation
(MainFragment.this.getActivity(), R.anim.push_bottom_in));
    MainFragment.this.vfRecomm.setOutAnimation(AnimationUtils.loadAnimation
(MainFragment.this.getActivity(), R.anim.push_bottom_out));
    MainFragment.this.vfRecomm.showPrevious();
    if (--recommPage ==0)
    {
        recommPage =totalPage;
    }
    textFooterMsg.setText(recommPage +"/" +totalPage);
}
```

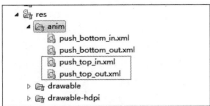

图 6.10 动画布局文件

android：duration＝"500"　设置动画时间为 0.5 秒。

android：fromYDelta＝"100％p"　设置 100％p 为一个相对值，大于 0 为下方，小于 0 为上方。此处表示动画开始时的位置在下方一个 ViewFlipper 的位置。

android：toYDelta＝"0"　动画结束时位置在当前 ViewFlipper 的位置。

android：fromAlpha＝"0.1"　开始时透明度为 0.1。

android：toAlpha="1.0"　结束时透明度为1。

```xml
<?xml version="1.0" encoding="utf-8"?>
<set xmlns:android="http://schemas.android.com/apk/res/android" >
    <translate
        android:duration="500"
        android:fromYDelta="100%p"
        android:toYDelta="0" />
    <alpha
        android:duration="500"
        android:fromAlpha="0.1"
        android:toAlpha="1.0" />
</set>
```

android：fromYDelta="0"　动画开始时当前位置。

android：toYDelta="－100%p"　动画结束时完全移动到上方。

android：fromAlpha="1.0"　开始时透明度为1。

android：toAlpha="0.1"　开始时透明度为0.1。

```xml
<?xml version="1.0" encoding="utf-8"?>
<set xmlns:android="http://schemas.android.com/apk/res/android" >
    <translate
        android:duration="500"
        android:fromYDelta="0"
        android:toYDelta="-100%p" />
    <alpha
        android:duration="500"
        android:fromAlpha="1.0"
        android:toAlpha="0.1" />
</set>
```

动态初始化布局时，绑定单击事件并进入商铺详情页。

```java
private void bindItem(ImageView iv, TextView tv)
{
    final HotsVendor hotsVendor =vendorList.get((recommPage -1) * 7 +count);
    tv.setText(hotsVendor.shopName);
    mImageFetcher.loadImage(hotsVendor.shopImage, iv);
    tv.setOnTouchListener(commonTouchListener);
    tv.setOnClickListener(dummyClickListener);
    iv.setOnTouchListener(commonTouchListener);
    iv.setOnClickListener(new OnClickListener()
    {
        public void onClick(View v)
        {
            if (!shouldPerformClick)
```

```
            {
                return;
            }
            toVendorDetail(hotsVendor);
        }
    });
    //计数器,每页初始化商铺数量为 7 个
    count++;
}
```

Bundle 作为 Intent 数据传输的对象,封装自定义 JavaBean 对象时,这个对象需要序列化才能传递。bundle.putSerializable("hotsVendor", hotsVendor);

```
public void toVendorDetail(HotsVendor hotsVendor)
{
    Intent intent = new Intent(getActivity(), VendorDetailActivity.class);
    Bundle bundle = new Bundle();
    bundle.putSerializable("hotsVendor", hotsVendor);
    intent.putExtras(bundle);
    startActivity(intent);
}
```

2) 全部商户

ShopsFragment 类图如图 6.11 所示。

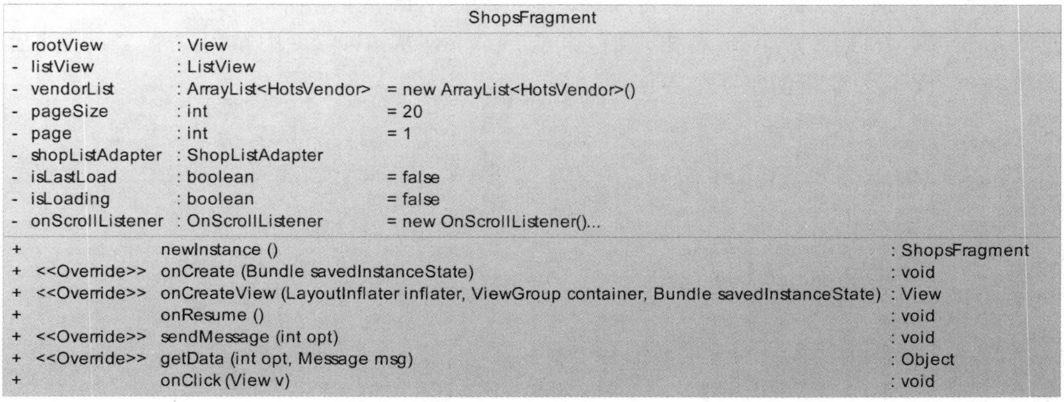

ShopsFragment		
- rootView	: View	
- listView	: ListView	
- vendorList	: ArrayList<HotsVendor>	= new ArrayList<HotsVendor>()
- pageSize	: int	= 20
- page	: int	= 1
- shopListAdapter	: ShopListAdapter	
- isLastLoad	: boolean	= false
- isLoading	: boolean	= false
- onScrollListener	: OnScrollListener	= new OnScrollListener()...
+ newInstance ()		: ShopsFragment
+ <<Override>> onCreate (Bundle savedInstanceState)		: void
+ <<Override>> onCreateView (LayoutInflater inflater, ViewGroup container, Bundle savedInstanceState)		: View
+ onResume ()		: void
+ <<Override>> sendMessage (int opt)		: void
+ <<Override>> getData (int opt, Message msg)		: Object
+ onClick (View v)		: void

图 6.11 ShopsFragment 类图

加载布局文件,并找到 ListView 组件。

```
rootView =inflater.inflate(R.layout.fragment_shops_main, container, false);
rootView.findViewById(R.id.image_top_layout_left).setOnClickListener(this);
```

创建 ListView 适配器 ShopListAdapter 类图,如图 6.12 所示。

```
shopListAdapter=new ShopListAdapter(mImageFetcher, getActivity(), vendorList);
```

```
listView.setAdapter(shopListAdapter);
```

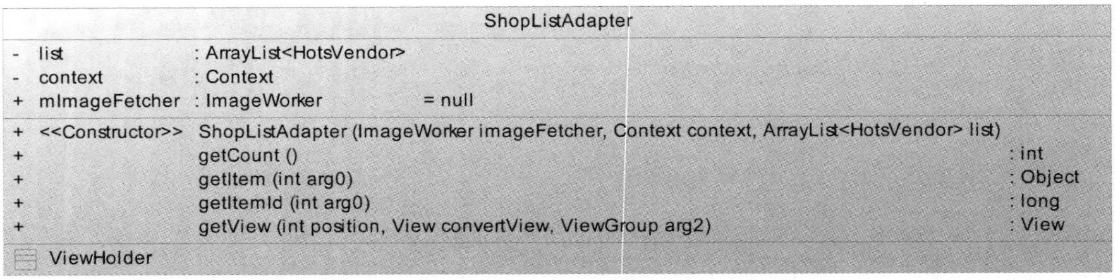

ShopListAdapter		
- list	: ArrayList<HotsVendor>	
- context	: Context	
+ mImageFetcher	: ImageWorker = null	
+ <<Constructor>> ShopListAdapter (ImageWorker imageFetcher, Context context, ArrayList<HotsVendor> list)		
+ getCount ()		: int
+ getItem (int arg0)		: Object
+ getItemId (int arg0)		: long
+ getView (int position, View convertView, ViewGroup arg2)		: View
ViewHolder		

图 6.12 适配器 ShopListAdapter 类图

下面介绍最重要的 Adapter.getView()方法。

这个方法获得指定位置显示的 View。官网解释为：Get a View that displays the data at the specified position in the data set. You can either create a View manually or inflate it from an xml layout file.

显示一个 View 就调用一次该方法。该方法涉及 ListView 性能的关键。该方法中有当前显示布局，Android 为其做过缓存机制。

ListView 中每个 item 都是通过 getView 返回并显示的。假如有很多个 item，那么重复创建这么多对象显然不合理。因此，Android 提供了 Recycler(回收站)，将没有正在显示的 item 放进 Recycle，然后在显示新视图时从 Recycle 中复用这个 View。

Recycler 的工作原理大致如下：

假设屏幕最多能看到 7 个 item，那么当第 1 个 item 滚出屏幕时，这个 item 的 View 进入 Recycle 中，第 8 个 item 要出现前，通过 getView 从 Recycle 中重用这个 View，然后设置数据，而不必重新创建一个 View。

```
@Override
public View getView(int position, View convertView, ViewGroup arg2)
{
    ViewHolder viewHolder;
    if (convertView ==null)
    {
        viewHolder =new ViewHolder();
        convertView =View.inflate(context, R.layout.adapter_shops_list_item, null);
        viewHolder.logo = (ImageView) convertView.findViewById(R.id.image_view_shops_picture_item_logo);
        viewHolder.name = (TextView) convertView.findViewById(R.id.image_view_shops_title_item_name);
        viewHolder.price = (TextView) convertView.findViewById(R.id.image_view_shops_item_price);
        viewHolder.shopDes = (TextView) convertView.findViewById(R.id.image_view_shops_item_desc);
```

```java
        convertView.setTag(viewHolder);
    } else
    {
        viewHolder = (ViewHolder) convertView.getTag();
    }
    final HotsVendor item = list.get(position);
    mImageFetcher.loadImage(item.shopImage, viewHolder.logo);
    viewHolder.name.setText(item.shopName);
    viewHolder.price.setText("人均消费：" + item.shopSpend);
    viewHolder.shopDes.setText(item.shopDesc);
    return convertView;
}
```

加载 xml 布局文件，内容如下：

```xml
<?xml version="1.0" encoding="utf-8"?>
<RelativeLayout xmlns:android="http://schemas.android.com/apk/res/android"
    android:id="@+id/layout_activity_item"
    android:layout_width="match_parent"
    android:layout_height="wrap_content"
    android:paddingBottom="5dp"
    android:paddingTop="5dp" >
    <ImageView
        android:id="@+id/image_view_shops_picture_item_logo"
        android:layout_width="90dp"
        android:layout_height="90dp"
        android:layout_centerVertical="true"
        android:contentDescription="@string/app_name"
        android:padding="5dp"
        android:scaleType="centerCrop"
        android:src="@drawable/member_pic" />
    <LinearLayout
        android:layout_width="match_parent"
        android:layout_height="wrap_content"
        android:layout_centerVertical="true"
        android:layout_marginLeft="10dp"
        android:layout_toRightOf="@id/image_view_shops_picture_item_logo"
        android:orientation="vertical" >
        <TextView
            android:id="@+id/image_view_shops_title_item_name"
            android:layout_width="match_parent"
            android:layout_height="wrap_content"
            android:singleLine="true"
            android:textColor="@color/black"
            android:textSize="16sp" />
```

```xml
<TextView
    android:id="@+id/image_view_shops_item_price"
    android:layout_width="match_parent"
    android:layout_height="wrap_content"
    android:singleLine="true"
    android:textColor="@color/grey"
    android:textSize="12sp" />
<TextView
    android:id="@+id/image_view_shops_item_desc"
    android:layout_width="match_parent"
    android:layout_height="wrap_content"
    android:textColor="@color/grey"
    android:textSize="12sp" />
    </LinearLayout>
</RelativeLayout>
```

因为每次滑动 List 的时候,都会调用 getView 方法,所以若不优化,图片或者条目加载过多将会导致内存溢出。

对于 ListView 的优化,一方面,对创建对象机制进行优化,不是每次执行 getView 方法都用 inflate 获得一个新的对象。要先检查这个 convertView 是不是已经存在。如果不存在,则创建一个对象;如果已经存在,就可以通过 getTag() 获得这个对象,进而减少对象数量的创建。另一方面,加大对图片的优化,比如使用软引用来降低内存的强制占用量,通过缓存提高图片访问速度。

谷歌 Android 开源组织提供的优化方法 com.me.demo.loadimg.ImageResizer.java 片段 1:

```java
public synchronized Bitmap decodeSampledBitmapFromFile(String filename, int reqWidth, int reqHeight)
{
    try
    {
        //First decode with inJustDecodeBounds=true to check dimensions
        final BitmapFactory.Options options = new BitmapFactory.Options();
        //若设置为 true,则 decode 时 Bitmap 的返回值为空,读取图片宽、高放在 Options 里
        options.inJustDecodeBounds = true;
        BitmapFactory.decodeFile(filename, options);
        //Calculate inSampleSize
        options.inSampleSize=calculateInSampleSize(options, reqWidth, reqHeight);
        //Decode bitmap with inSampleSize set
        options.inJustDecodeBounds = false;
        return BitmapFactory.decodeFile(filename, options);
    } catch (OutOfMemoryError e)
    {
```

```
        e.printStackTrace();
        return null;
    }
}
```

设置 OnItemClickListener 监听事件以及滑动监听事件,单击条目时,进入对应条目的商铺详情页面。

```java
listView.setOnItemClickListener(new AdapterView.OnItemClickListener()
{
    @Override
    public void onItemClick(AdapterView<?> parent, View v, int position, long id)
    {
        Intent intent = new Intent(ShopsFragment.this.getActivity(),
VendorDetailActivity.class);
        Bundle bundle = new Bundle();
        bundle.putSerializable("hotsVendor", vendorList.get(position));
        intent.putExtras(bundle);
        ShopsFragment.this.getActivity().startActivity(intent);
    }
});
listView.setOnScrollListener(onScrollListener);
```

onScrollListener 监听器。当滑动到底部时,若当前不是最后的数据,则请求服务器下一页的数据,如果数据已经请求完毕,则不再请求服务器数据。

```java
private OnScrollListener onScrollListener = new OnScrollListener()
{
    @Override
    public void onScrollStateChanged(AbsListView view, int scrollState)
    {
        if (scrollState ==AbsListView.OnScrollListener.SCROLL_STATE_FLING)
        {
            mImageFetcher.setPauseWork(true);
        } else
        {
            mImageFetcher.setPauseWork(false);
        }
         if (!isLoading && !isLastLoad && view.getLastVisiblePosition() ==
(view.getCount() -1))
            {
                isLoading = true;//锁上,防止重复操作
                sendMessage(OPT.GET_SHOPS);
            }
    }
```

```java
@Override
public void onScroll(AbsListView view, int firstVisibleItem, int visibleItemCount, int totalItemCount)
{
    //设置当前屏幕显示的起始 index 和结束 index
}
};
```

请求完数据,Adapter 刷新页面。

vendorList.addAll(newsList)导致数据发生变化,shopListAdapter.notifyDataSetChanged()方法刷新页面,当 List 的数据发生变化时,页面就会及时刷新。

isLastLoad=true;设置是否是最后一次数据。我们请求 pageSize 是每页返回的数据,如果数据已经足够了,那么返回的数据长度必然不小于 pageSize。当返回的数据小于 pageSize 时,则表明这是最后一部分数据,提示用户数据已经加载完毕。

```java
ArrayList<HotsVendor>newsList =(ArrayList<HotsVendor>) msg.obj;
if (newsList.size() <pageSize)
{
    Toast.makeText(getActivity(), "数据已加载完毕!", Toast.LENGTH_LONG).show();
    isLastLoad =true;
} else
{
    page++;
}
isLoading =false;//获取完毕,解开锁定
vendorList.addAll(newsList);
shopListAdapter.notifyDataSetChanged();
```

6.4.2 客户端和服务端交互

将数据封装到 JSON 对象中。

```java
param.put("act", "getShops");
param.put("pageSize", pageSize);
param.put("page", page);
```

服务端返回 JSON String 解析放到 JavaBean 文件 UserBean.java 中。

```java
package com.me.demo.bean;
import java.io.Serializable;
public class HotsVendor implements Serializable
{
    private static final long serialVersionUID =2282728744986789831L;
```

```
    public String shopId;
    public String shopAddress;
    public String shopName;
    public String shopTime;
    public String shopPhone;
    public String shopImage;
    public String shopLongitude;
    public String shopLatitude;
    public String shopType;
    public String shopDesc;
    public String shopSpend;
}
```

6.5 知识点回顾与技能扩展

6.5.1 知识点回顾

本章主要知识点如下：

（1）ViewFlipper 的理解与使用。

（2）动态组件的创建。

（3）ListView 的理解与使用。

（4）Adapter 的理解与使用。

6.5.2 技能扩展

1. 客户端图片

在开发过程中，一般都将经常要使用的图片缓存到本地 SD 卡中，以便下次直接引用而不需再次请求网络而耗费资源。如果自己开发的搜索功能里有涉及图片的显示，则可以直接利用 Map<String，SoftReference<Drawable>>，即软引用。

下面看看谷歌提供的源码是如何处理的，谷歌的图片处理方式全部放在 com.me.demo.loading 包中。

com.me.demo.loading.ImageWorker 的 loadImage（Object data，ImageView imageView）方法的处理流程是：bitmap = mImageCache.getBitmapFromMemCache（url），查看内存是否有这个图片，若查找到就利用 imageView.setImageBitmap（bitmap）设置图片。如果没有找到图片，并且图片也没有在队列中（cancelPotentialWork（url，imageView）），则启用后台程序去加载数据（AsyncDrawable）。

imageView.setTag（url）方法是 View 的方法，这个组件被标记后，可以通过 getTag 将其取出来。

imageView.setImageBitmap（bitmap）；设置该 ImageView 显示的图片。

BitmapWorkerTask 继承 AsyncTaskEx,而 AsyncTaskEx 是自定义的一个泛型,能够适当、简单地用于 UI 线程。这个类不需要操作线程(Thread)就可以完成后台操作将结果返回 UI。

执行 task.execute(url)后,就开始执行 doInBackground()方法。

```
Bitmap bitmap =null;
if (mImageCache !=null)
{
    bitmap =mImageCache.getBitmapFromMemCache(url);
}
imageView.setTag(url);
if (bitmap !=null)
{
    //Bitmap found in memory cache
    imageView.setImageBitmap(bitmap);
} else if (cancelPotentialWork(url, imageView))
{
    final BitmapWorkerTask task =new BitmapWorkerTask(imageView);
    final AsyncDrawable asyncDrawable =new AsyncDrawable(mContext.getResources(),
mLoadingBitmap, task);
    imageView.setImageDrawable(asyncDrawable);
    task.execute(url);
}
final BitmapWorkerTask task=new BitmapWorkerTask(imageView);
```

执行之前,先生成软引用 BitmapWorkerTask。

```
private static class AsyncDrawable extends BitmapDrawable
{
    private final WeakReference<BitmapWorkerTask>bitmapWorkerTaskReference;
    public AsyncDrawable(Resources res, Bitmap bitmap, BitmapWorkerTask
bitmapWorkerTask)
    {
        super(res, bitmap);
        bitmapWorkerTaskReference =new WeakReference<BitmapWorkerTask>(bitm
apWorkerTask);
    }
    public BitmapWorkerTask getBitmapWorkerTask()
    {
        return bitmapWorkerTaskReference.get();
    }
}
```

执行 doInBackground 方法。

mPauseWorkLock.wait();若展示线程被锁定,则等图片加载完毕解锁后再执行。

mImageCache.getBitmapFromDiskCache(dataString);在硬盘检查是否存在这张图片。

processBitmap(params[0])

```java
private class BitmapWorkerTask extends AsyncTaskEx<Object, Void, Bitmap>
{
    private Object data;
    private final WeakReference<ImageView> imageViewReference;
    public BitmapWorkerTask(ImageView imageView)
    {
        imageViewReference =new WeakReference<ImageView>(imageView);
    }
    /**
     * Background processing.
     */
    @Override
    protected Bitmap doInBackground(Object... params)
    {
        data =params[0];
        final String dataString =String.valueOf(data);
        Bitmap bitmap =null;
        synchronized (mPauseWorkLock)
        {
            while (mPauseWork && !isCancelled())
            {
                try
                {
                    mPauseWorkLock.wait();
                } catch (InterruptedException e)
                {
                }
            }
        }
        //SD卡获取图片
        if (mImageCache !=null && !isCancelled() && getAttachedImageView() != null && !mExitTasksEarly)
        {
            bitmap =mImageCache.getBitmapFromDiskCache(dataString);
        }
        //网络下载图片
        if (bitmap ==null && !isCancelled() && getAttachedImageView() !=null && !mExitTasksEarly)
```

```java
        {
            bitmap =processBitmap(params[0]);
        }
        //添加图片到缓存中
        if (bitmap !=null && mImageCache !=null)
        {
            mImageCache.addBitmapToCache(dataString, bitmap);
        }
        return bitmap;
    }
}
```

1) 从内存和硬盘获取图片

在 mImageCache. getBitmapFromDiskCache(dataString) 中, mDiskCache 是 com. me. demo. loadimg. DiskLruCache 的对象。

```java
public Bitmap getBitmapFromDiskCache(String data)
{
    if (mDiskCache !=null)
    {
        return mDiskCache.get(data);
    } else
    {
        return null;
    }
}
```

在方法 mDiskCache. get(data)中,查看 url 是否保存在对应关系的 mLinkedHashMap 中。若已保存,就从 Map 中获取该图片。如果没有保存,就在缓存目录中查看是否有该文件。

```java
private final Map<String, String>mLinkedHashMap =Collections.synchronizedMap(new LinkedHashMap<String, String>(INITIAL_CAPACITY, LOAD_FACTOR, true));

public Bitmap get(String key)
{
    synchronized (mLinkedHashMap)
    {
        final String file =mLinkedHashMap.get(key);
        try
        {
            if (file !=null)
            {
                if (BuildConfig.DEBUG)
                {
```

```java
                Log.d(TAG, "Disk cache hit");
            }
            return BitmapFactory.decodeFile(file);
        } else
        {
            final String existingFile = createFilePath(mCacheDir, key);
            if (new File(existingFile).exists())
            {
                put(key, existingFile);
                if (BuildConfig.DEBUG)
                {
                    Log.d(TAG, "Disk cache hit (existing file)");
                }
                return BitmapFactory.decodeFile(existingFile);
            }
        }
    } catch (OutOfMemoryError e)
    {
        flushCache();
        e.printStackTrace();
    }
    return null;
}
```

在 com. me. demo. loading. Util 中，图片缓存目录文件如下：

```java
public static File getExternalCacheDir(Context context)
{
    if (hasExternalCacheDir())
    {
        return context.getExternalCacheDir();
    } else
    {
        final String cacheDir = "/Android/data/" + context.getPackageName() + "/cache/";
        return new File(Environment.getExternalStorageDirectory().getPath() + cacheDir);
    }
}
```

2）从网络获取图片

processBitmap(params[0])方法从网络下载图片。

```java
private Bitmap processBitmap(String data)
{
    if (BuildConfig.DEBUG)
    {
        //Log.d(TAG, "processBitmap -" +data);
    }
    try
    {
        final File f =downloadBitmap(mContext, data);
        if (f !=null)
        {
            //Return a sampled down version
            return decodeSampledBitmapFromFile(f.toString(), mImageWidth,
mImageHeight);
        } else
        {
            return null;
        }
    } catch (OutOfMemoryError e)
    {
        e.printStackTrace();
        return null;
    }
}
```

DiskLruCache. getDiskCacheDir(context，HTTP_CACHE_DIR)；检查文件缓存地址是否存在。

DiskLruCache. openCache(context，cacheDir，HTTP_CACHE_SIZE)；创建可读可写的文件目录。

在 Manifest. xml 中申请对 SD 卡可读可写的权限。

```
<uses-permission android:name="android.permission.WRITE_EXTERNAL_STORAGE" />
<uses-permission android:name="android.permission.READ_EXTERNAL_STORAGE" />
```

new File(cache. createFilePath(urlString))；创建文件。

Utils. disableConnectionReuseIfNecessary()；检查是否有网络连接。

new BufferedInputStream(urlConnection. getInputStream()，Utils. IO_BUFFER_SIZE)；读取数据流。

new BufferedOutputStream（new FileOutputStream（cacheFile），Utils. IO_BUFFER_SIZE)；写入数据流。

```java
public static File downloadBitmap(Context context, String urlString)
{
```

```java
        final File cacheDir = DiskLruCache.getDiskCacheDir(context, HTTP_CACHE_
DIR);
        if (cacheDir == null)
        {
            return null;
        }
        final DiskLruCache cache = DiskLruCache.openCache(context, cacheDir, HTTP_
CACHE_SIZE);
        final File cacheFile = new File(cache.createFilePath(urlString));
        if (cache.containsKey(urlString))
        {
            if (BuildConfig.DEBUG)
            {
                Log.d(TAG, "downloadBitmap - found in http cache - " + urlString);
            }
            return cacheFile;
        }
        if (BuildConfig.DEBUG)
        {
            Log.d(TAG, "downloadBitmap - downloading - " + urlString);
        }
        Utils.disableConnectionReuseIfNecessary();
        HttpURLConnection urlConnection = null;
        BufferedOutputStream out = null;
        try
        {
            final URL url = new URL(urlString);
            urlConnection = (HttpURLConnection) url.openConnection();
            urlConnection.setRequestMethod("GET");
            urlConnection.setReadTimeout(6 * 1000);
            if (urlConnection.getResponseCode() == 200)
            {
                final InputStream in = new BufferedInputStream(urlConnection.
getInputStream(), Utils.IO_BUFFER_SIZE);
                out = new BufferedOutputStream(new FileOutputStream(cacheFile),
Utils.IO_BUFFER_SIZE);
                int b;
                while ((b = in.read()) != -1)
                {
                    out.write(b);
                }
                if (urlConnection != null)
                {
```

```java
                    urlConnection.disconnect();
            }
            if (out !=null)
            {
                try
                {
                    out.close();
                } catch (final IOException e)
                {
                    Log.e(TAG, "Error in downloadBitmap -" +e);
                }
            }
            return cacheFile;
        }
    } catch (final IOException e)
    {
        Log.e(TAG, "Error in downloadBitmap -" +e);
        if (urlConnection !=null)
        {
        urlConnection.disconnect();
        }
        if (out !=null)
        {
            try
            {
                out.close();
            } catch (final IOException e1)
            {
                Log.e(TAG, "Error in downloadBitmap -" +e1);
            }
        }
    }
    return null;
}
```

3）添加图片到缓存中

mImageCache. addBitmapToCache(dataString，bitmap)无论是从 SD 卡中获取还是从网络下载,都添加到缓存中。

```java
public void addBitmapToCache(String data, Bitmap bitmap)
{
    if (data ==null || bitmap ==null)
    {
```

```java
        return;
    }
    //添加到缓存
    if (mMemoryCache !=null && mMemoryCache.get(data) ==null)
    {
        mMemoryCache.put(data, bitmap);
    }
    //添加到 SD 卡中
    if (mDiskCache !=null && !mDiskCache.containsKey(data))
    {
        mDiskCache.put(data, bitmap);
    }
}
```

put(data，bitmap)写入到 SD 卡中，writeBitmapToFile(data, file)

```java
public void put(String key, Bitmap data)
{
    synchronized (mLinkedHashMap)
    {
        if (mLinkedHashMap.get(key) ==null)
        {
            try
            {
                final String file =createFilePath(mCacheDir, key);
                if (writeBitmapToFile(data, file))
                {
                    put(key, file);
                    flushCache();
                }
            } catch (final FileNotFoundException e)
            {
                Log.e(TAG, "Error in put: " +e.getMessage());
            } catch (final IOException e)
            {
                Log.e(TAG, "Error in put: " +e.getMessage());
            }
        }
    }
}
```

2. 图片处理

1) 将 bitmap 转换成 String

```java
public static byte[] bitmapToByte(String filePath)
{
    BufferedInputStream in;
    byte[] b = null;
    try
    {
        in = new BufferedInputStream(new FileInputStream(filePath));
        ByteArrayOutputStream out = new ByteArrayOutputStream(1024);
        byte[] temp = new byte[1024];
        int size = 0;
        while ((size = in.read(temp)) != -1)
        {
            out.write(temp, 0, size);
        }
        in.close();
        b = out.toByteArray();
    } catch (FileNotFoundException e)
    {
        e.printStackTrace();
    } catch (IOException e)
    {
        e.printStackTrace();
    }
    return b;
}
```

2)压缩 SD 卡的图片,返回 bitmap

```java
public static Bitmap getSmallBitmap(String filePath)
{
    final BitmapFactory.Options options = new BitmapFactory.Options();
    options.inJustDecodeBounds = true;
    BitmapFactory.decodeFile(filePath, options);
    //设置图片的大小
    options.inSampleSize = calculateInSampleSize(options, 480, 800);
    options.inJustDecodeBounds = false;
    return BitmapFactory.decodeFile(filePath, options);
}
```

3)计算图片的缩放值

```java
public static int calculateInSampleSize(BitmapFactory.Options options, int reqWidth, int reqHeight)
{
```

```java
        final int height = options.outHeight;
        final int width = options.outWidth;
        int inSampleSize = 1;
        if (height > reqHeight || width > reqWidth)
        {
            final int heightRatio = Math.round((float) height / (float) reqHeight);
            final int widthRatio = Math.round((float) width / (float) reqWidth);
            inSampleSize = heightRatio < widthRatio ? heightRatio : widthRatio;
        }
        return inSampleSize;
    }
```

4) 图片翻转 180°

```java
public static Bitmap postRotate(Context context, int drawable)
{
    Resources res = context.getResources();
    Bitmap img = BitmapFactory.decodeResource(res, drawable);
    Matrix matrix = new Matrix();
    matrix.postRotate(180); /* 翻转 180° */
    int width = img.getWidth();
    int height = img.getHeight();
    return Bitmap.createBitmap(img, 0, 0, width, height, matrix, true);
}
```

5) 画圆形图案

```java
public static Bitmap toRoundBitmap(Bitmap bitmap)
{
    int width = bitmap.getWidth();
    int height = bitmap.getHeight();
    float roundPx;
    float left, top, right, bottom, dst_left, dst_top, dst_right, dst_bottom;
    if (width <= height)
    {
        roundPx = width / 2;
        top = 0;
        bottom = width;
        left = 0;
        right = width;
        height = width;
        dst_left = 0;
        dst_top = 0;
        dst_right = width;
        dst_bottom = width;
```

```java
        } else
        {
            roundPx = height / 2;
            float clip = (width - height) / 2;
            left = clip;
            right = width - clip;
            top = 0;
            bottom = height;
            width = height;
            dst_left = 0;
            dst_top = 0;
            dst_right = height;
            dst_bottom = height;
        }
        Bitmap output = Bitmap.createBitmap(width, height, Config.ARGB_8888);
        Canvas canvas = new Canvas(output);
        final int color = 0xff424242;
        final Paint paint = new Paint();
        final Rect src = new Rect((int) left, (int) top, (int) right, (int) bottom);
        final Rect dst = new Rect((int) dst_left, (int) dst_top, (int) dst_right, (int) dst_bottom);
        final RectF rectF = new RectF(dst);
        paint.setAntiAlias(true);
        canvas.drawARGB(0, 0, 0, 0);
        paint.setColor(color);
        canvas.drawRoundRect(rectF, roundPx, roundPx, paint);
        paint.setXfermode(new PorterDuffXfermode(Mode.SRC_IN));
        canvas.drawBitmap(bitmap, src, dst, paint);
        return output;
    }
```

6）依据高度缩放

```java
public static Bitmap getScaleHImg(Bitmap bitmap, int newHeight)
{
    //图片源
    //Bitmap bm = BitmapFactory.decodeStream(getResources()
    //.openRawResource(id));
    if (bitmap == null)
    {
        return null;
    }
    //获得图片的宽、高
    int width = bitmap.getWidth();
```

```java
    int height =bitmap.getHeight();
    //设置想要的大小
    int newHeight1 =newHeight;
    int newWidth1 =width * newHeight1 / height;
    //计算缩放比例
    float scaleWidth =((float) newWidth1) / width;
    float scaleHeight =((float) newHeight1) / height;
    //取得想要缩放的Matrix参数
    Matrix matrix =new Matrix();
    matrix.postScale(scaleWidth, scaleHeight);
    //得到新的图片
    Bitmap newbm = Bitmap.createBitmap(bitmap, 0, 0, width, height, matrix, true);
    return newbm;
}
```

7) 依据宽度缩放

```java
public static Bitmap getScaleWImg(Bitmap bitmap, int newWidth)
{
    //图片源
    //Bitmap bm =BitmapFactory.decodeStream(getResources()
    //.openRawResource(id));
    if (bitmap ==null)
    {
        return null;
    }
    //获得图片的宽、高
    int width =bitmap.getWidth();
    int height =bitmap.getHeight();
    //设置想要的大小
    int newWidth1 =newWidth;
    int newHeight1 =height * newWidth1 / width;
    //计算缩放比例
    float scaleWidth =((float) newWidth1) / width;
    float scaleHeight =((float) newHeight1) / height;
    //取得想要缩放的Matrix参数
    Matrix matrix =new Matrix();
    matrix.postScale(scaleWidth, scaleHeight);
    //得到新的图片
    Bitmap newbm = Bitmap.createBitmap(bitmap, 0, 0, width, height, matrix, true);
    return newbm;
}
```

6.6 练　　习

（1）模仿你熟悉的电子商务 Android 应用实现静态的内容展示。
（2）实现虚拟商户信息的展示，效果如图 6.13 所示。
（3）实现客户端缓存的清理。

图 6.13　商户信息展示

第 7 章 支持用户基于 LBS 的应用

7.1 用户定位

在 Android 开发中地图和定位是很多软件不可或缺的内容,这些特色功能也给人们带来了很多方便。

Android 地图定位的选择有几种,谷歌地图、百度地图、腾讯 SOSO 地图、高德地图,它们各有优势与集成方式,需要根据需求去选择适合自己的第三方平台,这里就不一一介绍了。根据稳定性与实用性等方面的影响,本书主要讲解百度地图的集成开发。

使用百度 Android 定位 SDK,必须注册 GPS 和网络使用权限。要定位 SDK,需采用 GPS、基站、Wi-Fi 信号进行定位。当应用程序向定位 SDK 发起定位请求时,定位 SDK 会根据应用的定位因素(GPS、基站、Wi-Fi 信号)的实际情况(如是否开启 GPS,是否连接网络,是否有信号等)来生成相应定位依据进行定位。

用户可以设置满足自身需求的定位依据。若用户设置 GPS 优先,则优先使用 GPS 进行定位。如果 GPS 定位未打开或者没有可用的位置信息,且网络连接正常,定位 SDK 就会返回网络定位(即 Wi-Fi 与基站)的最优结果。为了使获得的网络定位结果更加精确,请打开手机的 Wi-Fi 开关。

7.1.1 LBS 与常见第三方地图服务简介

下面用百度地图和谷歌地图进行比较,总结如下:
百度地图的优势:
(1) 简洁明了的界面,符合大多数中国人的使用习惯。
(2) 地图的更新速度优于谷歌地图。
(3) 公交换乘的信息比谷歌地图更准确。
百度地图的劣势:
(1) 测距功能相当弱,只能进行直线测距。
(2) 广告太多。
(3) 没有直观的卫星地图。

谷歌地图的优势：
(1) 有地图和卫星图两种模式，能更清楚地了解地貌信息。
(2) 测距功能是根据路线轨迹计算，而不是直线距离，具有更加实际的参考意义。
(3) 全球的地图信息非常全面。
(4) 扩展功能丰富，如 Google Earth、3D 街景等。
(5) 提供了步行、公交、驾车等不同的通行方式，有不同的建议路线。
(6) 轨迹图更直观。

谷歌地图的劣势：
(1) 对中国小城市的地图信息不全面。
(2) 更新速度较慢，偶尔也会被其忽悠。
(3) 公交信息不全面。

7.1.2 在地图上找到自己

1. 如何在百度地图上找到自己

集成百度地图 Android 定位 SDK 步骤如下：
(1) 下载百度地图最新 SDK 集成包。
(2) 申请百度开发者账号，申请百度地图 SDK 的 key。
(3) 根据开发 API，在 Manifest.xml 填入以下内容。

```
<meta-data
        android:name="com.baidu.lbsapi.API_KEY"
        android:value="key" />//key:开发者申请的 key
```

(4) 初始化定位服务。

```
//百度定位服务
private LocationClient locationClient;
private MyLocationListener locationListener = new MyLocationListener();

protected void onCreate(android.os.Bundle savedInstanceState)
{
    super.onCreate(savedInstanceState);
    locationClient = new LocationClient(this);
    LocationClientOption option = new LocationClientOption();
    option.setLocationMode(LocationMode.Hight_Accuracy);//设置定位模式
    option.setCoorType("bd09ll");        //返回的定位结果是百度经纬度，默认值 gcj02
    option.setScanSpan(5000);            //设置发起定位请求的间隔时间为 5000ms
    option.setIsNeedAddress(true);       //返回的定位结果包含地址信息
    option.setNeedDeviceDirect(true);    //返回的定位结果包含手机机头的方向
    locationClient.setLocOption(option);
    locationClient.setLocOption(option);
```

```
            locationClient.registerLocationListener(locationListener);    //设置监听
            locationClient.start();
}
```

(5)百度监听器回调接口。

```
//百度定位监听器(回调函数)
public class MyLocationListener implements BDLocationListener
{
    @Override
    public void onReceiveLocation(BDLocation location)
    {
        if (location ==null)
            return;
        StringBuffer sb =new StringBuffer(256);
        sb.append("time : ");
        sb.append(location.getTime());
        sb.append("\nerror code : ");
        sb.append(location.getLocType());
        sb.append("\nlatitude : ");
        sb.append(location.getLatitude());
        sb.append("\nlontitude : ");
        sb.append(location.getLongitude());
        sb.append("\nradius : ");
        sb.append(location.getRadius());
        if (location.getLocType() ==BDLocation.TypeGpsLocation)
        {
            sb.append("\nspeed : ");
            sb.append(location.getSpeed());
            sb.append("\nsatellite : ");
            sb.append(location.getSatelliteNumber());
        } else if (location.getLocType() ==BDLocation.TypeNetWorkLocation)
        {
            sb.append("\naddr : ");
            sb.append(location.getAddrStr());
        }
    }
}
```

拿到经纬度坐标,定位基本完成。如果想知道周边商铺的数据,将经纬度坐标传到服务端解析计算周边商户的距离,就可以完成附件商户的搜索。

2. 如何在谷歌地图中找到自己

谷歌地图定位的开发步骤如下:
(1)得到认证指纹(MD5)。

（2）申请谷歌地图 api key。
（3）把申请到的 key 集成到 MapView 中。

```xml
<?xml version="1.0" encoding="utf-8"?>
<LinearLayout xmlns:android="http://schemas.android.com/apk/res/android"
    android:layout_width="fill_parent"
    android:layout_height="fill_parent"
    android:orientation="vertical" >
    <FrameLayout
        android:id="@+id/map_layout"
        android:layout_width="fill_parent"
        android:layout_height="fill_parent"
        android:orientation="vertical" >
        <com.google.android.maps.MapView
            android:id="@+id/map_view"
            android:layout_width="fill_parent"
            android:layout_height="fill_parent"
            android:apiKey="0Mg_koWoyZUhlluO4-i6-bq9WYMFbxKodZZMz2Q"
            android:clickable="true"
            android:enabled="true" />
        <LinearLayout
            android:layout_width="wrap_content"
            android:layout_height="wrap_content"
            android:layout_gravity="center"
            android:orientation="vertical"
            android:paddingBottom="105dip" >
            <TextView
                android:id="@+id/map_bubbleText"
                android:layout_width="wrap_content"
                android:layout_height="wrap_content"
                android:background="@drawable/location_tips"
                android:gravity="left|center"
                android:maxEms="12"
                android:paddingLeft="12dip"
                android:paddingRight="10dip"
                android:text="@string/load_tips"
                android:textColor="#cfcfcf"
                android:textSize="14sp" />
        </LinearLayout>
        <LinearLayout
            android:layout_width="wrap_content"
            android:layout_height="wrap_content"
            android:layout_gravity="center"
            android:orientation="vertical" >
```

```xml
        <ImageView
            android:id="@+id/point_image"
            android:layout_width="wrap_content"
            android:layout_height="wrap_content"
            android:layout_gravity="center"
            android:layout_marginBottom="30dip"
            android:src="@drawable/point_start" />
    </LinearLayout>
    </FrameLayout>
</LinearLayout>
```

（4）创建 MyLocationManager 类，主要用于管理经纬度获取方法的实现。

```java
package com.android.map;

import android.content.Context;
import android.location.Location;
import android.location.LocationListener;
import android.location.LocationManager;
import android.os.Bundle;
import android.util.Log;

public class MyLocationManager
{
    private final String TAG = "FzLocationManager";
    private static Context mContext;
    private LocationManager gpsLocationManager;
    private LocationManager networkLocationManager;
    private static final int MINTIME = 2000;
    private static final int MININSTANCE = 2;
    private static MyLocationManager instance;
    private Location lastLocation = null;
    private static LocationCallBack mCallback;
    public static void init(Context c, LocationCallBack callback)
    {
        mContext = c;
        mCallback = callback;
    }
    private MyLocationManager()
    {
        //GPS 定位
        gpsLocationManager = (LocationManager) mContext.getSystemService(Context.LOCATION_SERVICE);
        Location gpsLocation = gpsLocationManager.getLastKnownLocation(LocationManager.GPS_PROVIDER);
```

```java
        gpsLocationManager.requestLocationUpdates(LocationManager.GPS_PROVIDER,
MINTIME, MININSTANCE, locationListener);
        //基站定位
        networkLocationManager = (LocationManager) mContext.getSystemService
(Context.LOCATION_SERVICE);
        Location networkLocation =gpsLocationManager.getLastKnownLocation
(LocationManager.GPS_PROVIDER);
        networkLocationManager.requestLocationUpdates(LocationManager.
NETWORK_PROVIDER, MINTIME, MININSTANCE, locationListener);
    }
    public static MyLocationManager getInstance()
    {
        if (null ==instance)
        {
            instance =new MyLocationManager();
        }
        return instance;
    }
    private void updateLocation(Location location)
    {
        lastLocation =location;
        mCallback.onCurrentLocation(location);
    }
    private final LocationListener locationListener =new LocationListener()
    {
        public void onStatusChanged(String provider, int status, Bundle extras)
        {
        }
        public void onProviderEnabled(String provider)
        {
        }
        public void onProviderDisabled(String provider)
        {
        }
        public void onLocationChanged(Location location)
        {
            Log.d(TAG, "onLocationChanged");
            updateLocation(location);
        }
    };
    public Location getMyLocation()
    {
        return lastLocation;
    }
```

```java
    private static int ENOUGH_LONG =1000 * 60;
    public interface LocationCallBack
    {
        /**
         * 当前位置
         *
         * @param location
         */
        void onCurrentLocation(Location location);
    }
    public void destoryLocationManager()
    {
        Log.d(TAG, "destoryLocationManager");
        gpsLocationManager.removeUpdates(locationListener);
        networkLocationManager.removeUpdates(locationListener);
    }
}
```

（5）创建 MyMapOverlay 抽象类，并继承 Overlay。创建抽象方法 changePoint (GeoPoint newPoint, int type)用于回调重新获取的 GeoPoint。重新定位地图，并获取地址信息。

```java
import android.view.MotionEvent;
import com.google.android.maps.GeoPoint;
import com.google.android.maps.MapView;
import com.google.android.maps.Overlay;
//覆盖整个地图捕捉触控事件的 OverLay
public abstract class MyMapOverlay extends Overlay
{
    private int point_X;
    private int point_Y;
    private GeoPoint newPoint;
    public MyMapOverlay(int x, int y)
    {
        point_X =x;
        point_Y =y;
    }
    boolean flagMove =false;
    //触控屏幕移动地图,重新根据屏幕中心点获取该点经纬度
    @Override
    public boolean onTouchEvent(MotionEvent event, MapView mapView)
    {
        System.out.println("X->" +event.getX() +":" +point_X);
        System.out.println("Y->" +event.getY() +":" +point_Y);
        if (event.getAction() ==MotionEvent.ACTION_DOWN)
```

```
            {
                changePoint(newPoint, 1);
            } else if (event.getAction() ==MotionEvent.ACTION_UP)
            {
                newPoint =mapView.getProjection().fromPixels(point_X, point_Y);
                changePoint(newPoint, 2);
            }
            return false;
        }
        public abstract void changePoint(GeoPoint newPoint, int type);
    }
```

（6）MyMapActivity 继承 MapActivity 类并实现经纬度获取回调接口 LocationCallBack。

```
package com.android.googlemap;

import java.io.IOException;
import java.util.List;
import java.util.Locale;
import android.graphics.Rect;
import android.location.Address;
import android.location.Geocoder;
import android.location.Location;
import android.os.Bundle;
import android.os.Handler;
import android.os.Message;
import android.view.View;
import android.view.Window;
import android.widget.TextView;
import com.android.map.MyLocationManager;
import com.android.map.MyLocationManager.LocationCallBack;
import com.android.map.MyMapOverlay;
import com.google.android.maps.GeoPoint;
import com.google.android.maps.MapActivity;
import com.google.android.maps.MapController;
import com.google.android.maps.MapView;
import com.google.android.maps.Overlay;

public class MyMapActivity extends MapActivity implements LocationCallBack
{
    private MapView mapView;
    private MapController mMapCtrl;
    private MyLocationManager myLocation;
    private List<Overlay>mapOverlays;
    public GeoPoint locPoint;
```

```java
private MyMapOverlay mOverlay;
private TextView desText;
private String lost_tips;
private int point_X;
private int point_Y;
private int statusBarHeight;
public final int MSG_VIEW_LONGPRESS =10001;
public final int MSG_VIEW_ADDRESSNAME =10002;
public final int MSG_GONE_ADDRESSNAME =10003;
@Override
public void onCreate(Bundle savedInstanceState)
{
    super.onCreate(savedInstanceState);
    requestWindowFeature(Window.FEATURE_NO_TITLE);
    setContentView(R.layout.main);
    mapView = (MapView) findViewById(R.id.map_view);
    desText = (TextView) this.findViewById(R.id.map_bubbleText);
    lost_tips =getResources().getString(R.string.load_tips);
    mapView.setBuiltInZoomControls(true);
    mapView.setClickable(true);
    mMapCtrl =mapView.getController();
    point_X =this.getWindowManager().getDefaultDisplay().getWidth() / 2;
    point_Y =this.getWindowManager().getDefaultDisplay().getHeight() / 2;
    mOverlay =new MyMapOverlay(point_X, point_Y)
    {
        @Override
        public void changePoint(GeoPoint newPoint, int type)
        {
            if (type ==1)
            {
                mHandler.sendEmptyMessage(MSG_GONE_ADDRESSNAME);
            } else
            {
                locPoint =newPoint;
                mHandler.sendEmptyMessage(MSG_VIEW_LONGPRESS);
            }
        }
    };
    mapOverlays =mapView.getOverlays();
    if (mapOverlays.size() >0)
    {
        mapOverlays.clear();
    }
    mapOverlays.add(mOverlay);
```

```java
        mMapCtrl.setZoom(12);
    MyLocationManager.init(MyMapActivity.this.getApplicationContext(),
MyMapActivity.this);
        myLocation = MyLocationManager.getInstance();
    }
    @Override
    protected void onResume()
    {
        super.onResume();
    }
    @Override
    protected void onPause()
    {
        super.onPause();
    }
    @Override
    protected boolean isRouteDisplayed()
    {
        return false;
    }
    public void onCurrentLocation(Location location)
    {
        if (locPoint == null)
        {
            locPoint = new GeoPoint((int) (location.getLatitude() * 1E6), (int) (location.getLongitude() * 1E6));
            mHandler.sendEmptyMessage(MSG_VIEW_LONGPRESS);
        }
    }
    public void changePoint(GeoPoint locPoint)
    {
    }
    /**
     * 通过经、纬度获取地址
     * @param point
     * @return
     */
    private String getLocationAddress(GeoPoint point)
    {
        String add = "";
        Geocoder geoCoder = new Geocoder(getBaseContext(), Locale.getDefault());
        try
        {
            List<Address> addresses = geoCoder.getFromLocation(point.
```

```java
        getLatitudeE6() / 1E6, point.getLongitudeE6() / 1E6, 1);
            Address address = addresses.get(0);
            int maxLine = address.getMaxAddressLineIndex();
            if (maxLine >= 2)
            {
                add = address.getAddressLine(1) + address.getAddressLine(2);
            } else
            {
                add = address.getAddressLine(1);
            }
        } catch (IOException e)
        {
            add = "";
            e.printStackTrace();
        }
        return add;
    }
    /**
     * 用线程异步获取
     */
    Runnable getAddressName = new Runnable()
    {
        public void run()
        {
            String addressName = "";
            while (true)
            {
                addressName = getLocationAddress(locPoint);
                if (!"".equals(addressName))
                {
                    break;
                }
            }
            Message msg = new Message();
            msg.what = MSG_VIEW_ADDRESSNAME;
            msg.obj = addressName;
            mHandler.sendMessage(msg);
        }
    };
    private Handler mHandler = new Handler()
    {
        @Override
        public void handleMessage(Message msg)
        {
```

```java
            switch (msg.what)
            {
            case MSG_VIEW_LONGPRESS://处理长按时间返回位置信息
            {
                if (null == locPoint)
                    return;
                new Thread(getAddressName).start();
                desText.setVisibility(View.VISIBLE);
                desText.setText(lost_tips);
                mMapCtrl.animateTo(locPoint);
                mapView.invalidate();
            }
                break;
            case MSG_VIEW_ADDRESSNAME:
                desText.setText((String) msg.obj);
                desText.setVisibility(View.VISIBLE);
                if (statusBarHeight == 0)
                {
                    Rect frame = new Rect();    getWindow().getDecorView().getWindowVisibleDisplayFrame(frame);
                    statusBarHeight = frame.top;
                    point_Y -= statusBarHeight / 2;
                }
                break;
            case MSG_GONE_ADDRESSNAME:
                desText.setVisibility(View.GONE);
                break;
            }
        }
    };
    //关闭程序也关闭定位
    @Override
    protected void onDestroy()
    {
        super.onDestroy();
        myLocation.destoryLocationManager();
    }
}
```

7.2 摇 一 摇

7.2.1 摇一摇功能的实现

ShakeFragment 类图.java 如图 7.1 所示。

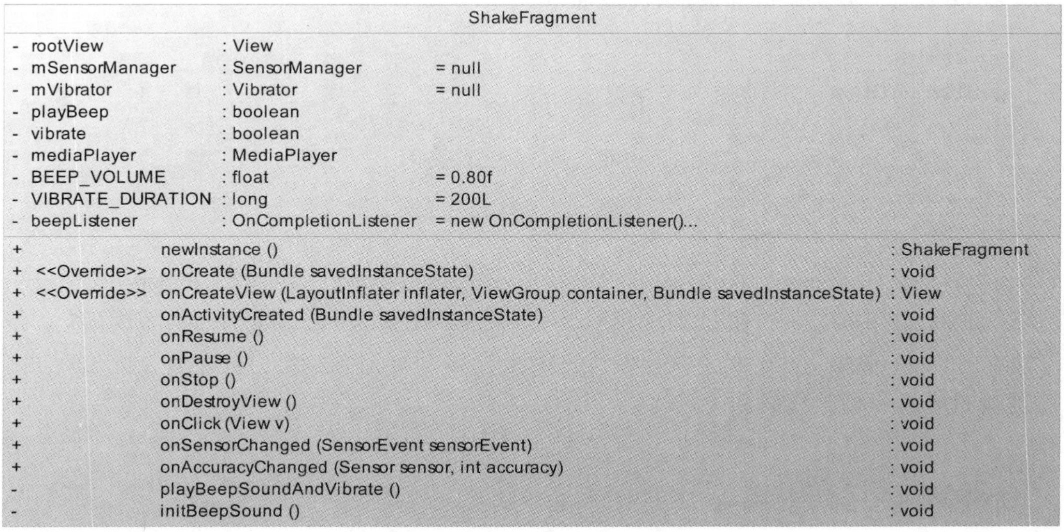

ShakeFragment			
- rootView	: View		
- mSensorManager	: SensorManager	= null	
- mVibrator	: Vibrator	= null	
- playBeep	: boolean		
- vibrate	: boolean		
- mediaPlayer	: MediaPlayer		
- BEEP_VOLUME	: float	= 0.80f	
- VIBRATE_DURATION	: long	= 200L	
- beepListener	: OnCompletionListener	= new OnCompletionListener()...	
+	newInstance ()		: ShakeFragment
+ <<Override>>	onCreate (Bundle savedInstanceState)		: void
+ <<Override>>	onCreateView (LayoutInflater inflater, ViewGroup container, Bundle savedInstanceState)		: View
+	onActivityCreated (Bundle savedInstanceState)		: void
+	onResume ()		: void
+	onPause ()		: void
+	onStop ()		: void
+	onDestroyView ()		: void
+	onClick (View v)		: void
+	onSensorChanged (SensorEvent sensorEvent)		: void
+	onAccuracyChanged (Sensor sensor, int accuracy)		: void
-	playBeepSoundAndVibrate ()		: void
-	initBeepSound ()		: void

图 7.1　ShakeFragment. java 类图

在 onCreateView 中获得传感器管理类 SensorManager，代码如下：

```
//获得传感器管理类
mSensorManager = (SensorManager) getActivity().getSystemService(Activity.SENSOR_SERVICE);
//获取手机震动对象
mVibrator = (Vibrator) getActivity().getSystemService(Activity.VIBRATOR_SERVICE);
```

在 onResume 中获取手机三个轴上的加速力，并初始化音频。代码如下：

```
//注册三轴传感器来监听其加速度
mSensorManager.registerListener(this, mSensorManager.getDefaultSensor(Sensor.TYPE_ACCELEROMETER), SensorManager.SENSOR_DELAY_NORMAL);
//播放音频
playBeep = true;
//获得音频管理器
AudioManager audioService = (AudioManager) getActivity().getSystemService(Activity.AUDIO_SERVICE);
if (audioService.getRingerMode() != AudioManager.RINGER_MODE_NORMAL)
//设置音频模式为普通
{
    playBeep = false;
}
//初始化音频
initBeepSound();
//开启震动
vibrate = true;
```

在 onStop() 方法中解除注册,代码如下:

```java
@Override
public void onPause()
{
    mSensorManager.unregisterListener(this);
    super.onPause();
}
```

编写 ShakeFragment implements SensorEventListener 接口,并重写 onSensorChanged (SensorEvent sensorEvent) 和 onAccuracyChanged(Sensor sensor, int accuracy)方法。

在 onSensorChanged 方法中监听到三轴坐标移动绝对值大于 14 这个边界值时,执行震动和播放音频。代码如下:

```java
@Override
public void onSensorChanged(SensorEvent sensorEvent)
{
    int sensorType =sensorEvent.sensor.getType();
    float[] values =sensorEvent.values;
    if (sensorType ==Sensor.TYPE_ACCELEROMETER)
    {
        if (Math.abs(values[0]) >14 || Math.abs(values[1]) >14 || Math.abs(values[2]) >14)
        {
            playBeepSoundAndVibrate();
        }
    }
}
@Override
public void onAccuracyChanged(Sensor sensor, int accuracy)
{}
```

播放音频,实现手机震动方法,同时请求服务端数据。代码如下:

```java
private void playBeepSoundAndVibrate()
{
    if (playBeep && mediaPlayer !=null)
    {
        mediaPlayer.start();
    }
    if (vibrate)
    {
        mVibrator.vibrate(VIBRATE_DURATION);
    }
    Intent intent =new Intent(getActivity(), SearchShopResultActivity.class);
    intent.putExtra("type", String.valueOf(new Random().nextInt(2) +1));
```

```java
        intent.putExtra("range", String.valueOf(3));
        intent.putExtra("keyword", "");
        getActivity().startActivity(intent);
}
/**
 * 初始化音频播放器
 */
private void initBeepSound()
{
    if (playBeep && mediaPlayer ==null)
    {
        getActivity().setVolumeControlStream(AudioManager.STREAM_MUSIC);
        mediaPlayer =new MediaPlayer();
        mediaPlayer.setAudioStreamType(AudioManager.STREAM_MUSIC);
        mediaPlayer.setOnCompletionListener(beepListener);

        AssetFileDescriptor file =getResources().openRawResourceFd(R.raw.shake_sound_male);
        try
        {
            mediaPlayer.setDataSource(file.getFileDescriptor(), file.getStartOffset(), file.getLength());
            file.close();
            mediaPlayer.setVolume(BEEP_VOLUME, BEEP_VOLUME);
            mediaPlayer.prepare();
        } catch (IOException e)
        {
            mediaPlayer =null;
        }
    }
}
private final OnCompletionListener beepListener =new OnCompletionListener()
{
    public void onCompletion(MediaPlayer mediaPlayer)
    {
        mediaPlayer.seekTo(0);
    }
};
```

7.2.2 传感器介绍

要知道手机中有多少种传感器，可以通过传感器管理类 SensorManager 获取。代码如下：

```java
//从系统服务中获得传感器管理器
SensorManager sm = (SensorManager) getActivity().getSystemService(Context.SENSOR_SERVICE);
//从传感器管理器中获得全部传感器的列表
List<Sensor> allSensors = sm.getSensorList(Sensor.TYPE_ALL);
//显示有多少个传感器
TextView tv = new TextView(getActivity());
tv.setText("经检测该手机有" + allSensors.size() + "个传感器,分别是:\n");
//显示每个传感器的具体信息
for (Sensor s : allSensors)
{
    String tempString = "\n" + "  设备名称:" + s.getName() + "\n" + "  设备版本:" + s.getVersion() + "\n" + "  供应商:" + s.getVendor() + "\n";
    System.out.println(tempString);
}
```

传感器监听接口实现 SensorEventListener。在重写的 onSensorChanged 方法中,当传感器坐标位置发生改变时,其临界值 14 被超过后,我们判断用户在摇动手机,那么就开始播放音乐,手机震动,并完成用户的各种操作。

```java
@Override
public void onSensorChanged(SensorEvent sensorEvent)
{
    int sensorType = sensorEvent.sensor.getType();
    float[] values = sensorEvent.values;
    if (sensorType == Sensor.TYPE_ACCELEROMETER)
    {
        if (Math.abs(values[0]) > 14 || Math.abs(values[1]) > 14 || Math.abs(values[2]) > 14)
        {
            playBeepSoundAndVibrate();
        }
    }
}
@Override
public void onAccuracyChanged(Sensor sensor, int accuracy)
{
}
```

7.3 知识点回顾

本章主要知识点如下:
(1) 传感器的管理类 SensorManager 以及监听事件的理解与使用。

（2）音频视频播放类 MediaPlayer 的理解与使用。

7.4 练 习

（1）用百度地图显示自己的位置。

用百度地图显示自己的位置，分为两步。第一步：定位；第二步：显示到百度地图上。定位已经在前面章节中介绍过，这里简单讲解怎么显示到百度地图上。关键代码如下：

```java
/**
 * 实现实位回调监听
 */
public class MyLocation implements BDLocationListener
{
    @Override
    public void onReceiveLocation(BDLocation location)
    {
        MeConfig.bdLocation = location;
        MyLocationData locData = new MyLocationData.Builder().accuracy(location.getRadius()).direction(100).latitude(location.getLatitude()).longitude(location.getLongitude()).build();
        mBaiduMap.setMyLocationData(locData);
    }
}
```

mBaiduMap.setMyLocationData(MyLocationData)这个方法就是将定位到的对象显示到百度地图上。

（2）摇出周边商户的信息。

第 8 章 用户搜索与结果展示

8.1 用户搜索功能总体设计

用户搜索时序如图 8.1 所示。

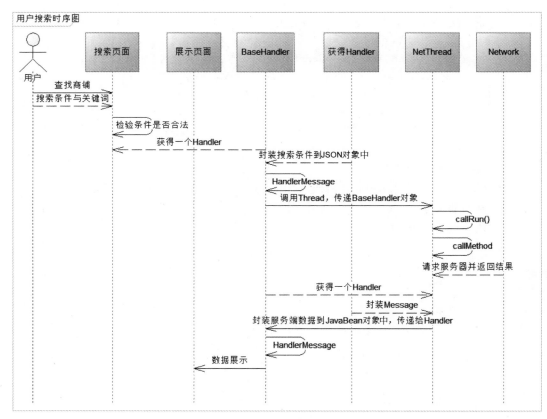

图 8.1 用户搜索时序

用户搜索流程如图 8.2 所示。

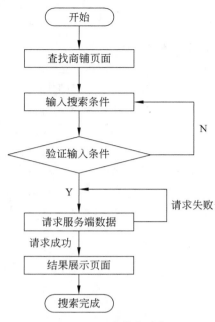

图 8.2 用户搜索流程

8.2 用户搜索功能知识点详解

1. PopupWindow 介绍

PopupWindow 所在包：

android.widget.PopupWindow

功能角色：PopupWindow 可以用来装载一些信息或 View，它可以悬浮在当前活动窗口上，并且不干扰用户对背后窗口的操作。

2. PopupWindow 的几个重要方法

（1）static View inflate(Context context, int resource, View Grouproot)。
加载一个布局并加载到返回的 View 中。
（2）PopupWindow(View contentView, int width, int height, boolean focusable)。
创建一个 PopupWindow 对象。
contentView：包含 PopupWindow 布局的 View。
width：PopupWindow 的宽度。
height：PopupWindow 的高度。
focusable：PopupWindow 是否聚焦。

```
popupWindow_view =View.inflate(getActivity(), layoutId, null);
```

```
popupWindow =new PopupWindow(popupWindow_view, ViewGroup.LayoutParams.MATCH_
PARENT, ViewGroup.LayoutParams.MATCH_PARENT, true);
```

(3) setInputMethodMode(int mode)。

设置 PopupWindow 的输入法模式：避免输入法弹出效果影响到 PopupWindow。

PopupWindow.INPUT_METHOD_NOT_NEEDED：不允许输入法。

PopupWindow.INPUT_METHOD_NEEDED：允许输入法。

PopupWindow.INPUT_METHOD_FROM_FOCUSABLE：根据是否可以有焦点确定输入法。

(4) showAtLocation(View parent, int gravity, int x, int y)。

(5) showAsDropDown(View anchor)、showAsDropDown(View anchor, int xoff, int yoff)、showAsDropDown(View anchor, int xoff, int yoff, int gravity)。

显示在 View 上方：

```
popupWindow.showAtLocation(v, Gravity.NO_GRAVITY, location[0], location[1]-
popupWindow.getHeight());
```

显示在 View 下方：

```
popupWindow.showAsDropDown(v);
```

显示在 View 左方：

```
popupWindow.showAtLocation(v, Gravity.NO_GRAVITY, location[0]-popupWindow.
getWidth(), location[1]);
```

显示在 View 右方：

```
popupWindow.showAtLocation(v, Gravity.NO_GRAVITY, location[0]+v.getWidth(),
location[1]);
```

注意：只有当 View 加载完成之后才能显示 PopupWindow。如果 View 没有加载完成就加载 PopupWindow，则不会成功。

判断 View 是否加载完成可以判断其宽度是否为 0，若不为 0，则加载完成。然后再加载 PopupWindow，这里可以用 Handler 来实现。

SearchRulesFragment.java 中 PopupWindow 相关的使用代码如下：

```
getPopupWindow(R.layout.popwindow_choice_types);
popupWindow.setInputMethodMode(PopupWindow.INPUT_METHOD_NOT_NEEDED);
popupWindow.showAtLocation(v, Gravity.TOP, 0, 0);

private void getPopupWindow(int layoutId)
{
    if (null !=popupWindow)
    {
        popupWindow.dismiss();
        popupWindow =null;
```

```
        }
        initPopWindow(layoutId);
}

private void initPopWindow(int layoutId)
    {
        popupWindow_view =View.inflate(getActivity(), layoutId, null);
        popupWindow =new PopupWindow(popupWindow_view, ViewGroup.LayoutParams.
MATCH_PARENT, ViewGroup.LayoutParams.MATCH_PARENT, true);
        switch (layoutId)
        {
        case R.layout.popwindow_choice_types:        popupWindow_view.
findViewById(R.id.tv_type_one).setOnClickListener(this);
            popupWindow_view.findViewById(R.id.tv_type_two).setOnClickListener
(this);
            popupWindow_view.findViewById(R.id.tv_type_all).setOnClickListener
(this);
            popupWindow_view.findViewById(R.id.tv_type_all_cancel).setOnClickListener
(this);
                break;
            case R.layout.popwindow_choice_distance:
    popupWindow_view.findViewById(R.id.tv_distance_one).setOnClickListener
(this);
            popupWindow_view.findViewById(R.id.tv_distance_two).setOnClickListener
(this);
            popupWindow_view.findViewById(R.id.tv_distance_all).setOnClickListener
(this);
            popupWindow_view.findViewById(R.id.tv_distance_cancel).setOnClickListener
(this);
                break;
        }
        ColorDrawable dw =new ColorDrawable(-00000);
        popupWindow.setBackgroundDrawable(dw);
        popupWindow.update();
}
```

(6) dismiss()。

当选择完成,PopupWindow 要消失时,调用 dismiss()方法。

```
popupWindow.dismiss();
```

8.3 用户搜索的实现

1. 用户搜索界面效果及实现

查找商户效果,如图 8.3、图 8.4、图 8.5 所示。

图 8.3　查找商户界面　　　　　图 8.4　选择商户类型界面

图 8.5　选择商户区域界面

　　底部布局与前面章节的布局大同小异，这里不再详述。下面主要讲解关于 Popwindow 的使用。

　　先看两个 Popwindow 的 XML 代码。

　　选择商铺类型的 Popwindow 界面布局文件 popwindow_choice_types.xml 的代码

如下：

```xml
<?xml version="1.0" encoding="utf-8"?>
<RelativeLayout xmlns:android="http://schemas.android.com/apk/res/android"
    android:layout_width="match_parent"
    android:layout_height="match_parent"
    android:background="@color/e0757575" >
    <RelativeLayout
        android:layout_width="match_parent"
        android:layout_height="240dp"
        android:layout_alignParentBottom="true"
        android:layout_marginLeft="10dp"
        android:layout_marginRight="10dp"
        android:background="@drawable/mejust_editbox_selector"
        android:paddingLeft="20dp"
        android:paddingRight="20dp" >
        <TextView
            android:id="@+id/tv_tips_choice_types"
            android:layout_width="match_parent"
            android:layout_height="wrap_content"
            android:gravity="center_horizontal|center_vertical"
            android:text="@string/tips_choice_types" />
        <TextView
            android:id="@+id/tv_type_one"
            android:layout_width="match_parent"
            android:layout_height="40dp"
            android:layout_below="@id/tv_tips_choice_types"
            android:layout_marginTop="5dp"
            android:background="@drawable/mejust_editbox_selector"
            android:gravity="center"
            android:text="@string/tips_choice_meal"
            android:textColor="@color/b3d3"
            android:textSize="16sp" >
        </TextView>
        <TextView
            android:id="@+id/tv_type_two"
            android:layout_width="match_parent"
            android:layout_height="40dp"
            android:layout_below="@id/tv_type_one"
            android:layout_marginTop="2dp"
            android:background="@drawable/mejust_editbox_selector"
            android:gravity="center"
            android:text="@string/tips_choice_media"
```

```xml
            android:textColor="@color/b3d3"
            android:textSize="16sp" >
        </TextView>
        <TextView
            android:id="@+id/tv_type_all"
            android:layout_width="match_parent"
            android:layout_height="40dp"
            android:layout_below="@id/tv_type_two"
            android:layout_marginTop="2dp"
            android:background="@drawable/mejust_editbox_selector"
            android:gravity="center"
            android:text="@string/all_types"
            android:textColor="@color/b3d3"
            android:textSize="16sp" >
        </TextView>
        <TextView
            android:id="@+id/tv_type_all_cancel"
            android:layout_width="match_parent"
            android:layout_height="40dp"
            android:layout_below="@id/tv_type_all"
            android:layout_marginTop="20dp"
            android:background="@drawable/mejust_editbox_selector"
            android:gravity="center"
            android:text="@string/cancle"
            android:textColor="@color/b3d3"
            android:textSize="16sp"
            android:textStyle="bold" >
        </TextView>
    </RelativeLayout>
</RelativeLayout>
```

选择距离的 Popwindow 界面布局文件，popwindow_choice_distance.xml 的代码如下：

```xml
<?xml version="1.0" encoding="utf-8"?>
<RelativeLayout xmlns:android="http://schemas.android.com/apk/res/android"
    android:layout_width="match_parent"
    android:layout_height="match_parent"
    android:background="@color/e0757575" >
    <RelativeLayout
        android:layout_width="match_parent"
        android:layout_height="240dp"
        android:layout_alignParentBottom="true"
        android:layout_marginLeft="10dp"
```

```xml
        android:layout_marginRight="10dp"
        android:background="@drawable/mejust_editbox_selector"
        android:paddingLeft="20dp"
        android:paddingRight="20dp" >
        <TextView
            android:id="@+id/tv_tips_choice_types"
            android:layout_width="match_parent"
            android:layout_height="wrap_content"
            android:gravity="center_horizontal|center_vertical"
            android:text="@string/tips_choice_dist" />
        <TextView
            android:id="@+id/tv_distance_one"
            android:layout_width="match_parent"
            android:layout_height="40dp"
            android:layout_below="@id/tv_tips_choice_types"
            android:layout_marginTop="5dp"
            android:background="@drawable/mejust_editbox_selector"
            android:gravity="center"
            android:text="@string/five_h_miters"
            android:textColor="@color/b3d3"
            android:textSize="16sp" >
        </TextView>
        <TextView
            android:id="@+id/tv_distance_two"
            android:layout_width="match_parent"
            android:layout_height="40dp"
            android:layout_below="@id/tv_distance_one"
            android:layout_marginTop="2dp"
            android:background="@drawable/mejust_editbox_selector"
            android:gravity="center"
            android:text="@string/one_t_miters"
            android:textColor="@color/b3d3"
            android:textSize="16sp" >
        </TextView>
        <TextView
            android:id="@+id/tv_distance_all"
            android:layout_width="match_parent"
            android:layout_height="40dp"
            android:layout_below="@id/tv_distance_two"
            android:layout_marginTop="2dp"
            android:background="@drawable/mejust_editbox_selector"
            android:gravity="center"
            android:text="@string/all_dist"
            android:textColor="@color/b3d3"
```

```xml
                android:textSize="16sp" >
        </TextView>
        <TextView
                android:id="@+id/tv_distance_cancel"
                android:layout_width="match_parent"
                android:layout_height="40dp"
                android:layout_below="@id/tv_distance_all"
                android:layout_marginTop="20dp"
                android:background="@drawable/mejust_editbox_selector"
                android:gravity="center"
                android:text="@string/cancle"
                android:textColor="@color/b3d3"
                android:textSize="16sp"
                android:textStyle="bold" >
        </TextView>
    </RelativeLayout>
</RelativeLayout>
```

从两个布局文件可以看出，Popwindow 弹出窗口的布局与一般的布局文件差别不大。这两个文件是全屏的 Popwindow，只是背景设置为半透明的蒙版模式，只需设置组件的背景色包含透明通道，即 android：background＝"@color/e0757575"，至于要实现什么样的效果，可根据项目需要进行改变。

还有一种非全屏的 Popwindow，它布局时只需要布局弹出的部分界面。需要注意的是展示 Popwindow 的几种方式，这在代码实现时再讲解。

2. 用户搜索功能的流程控制

创建 SearchRulesFragment 类图，如图 8.6 所示。

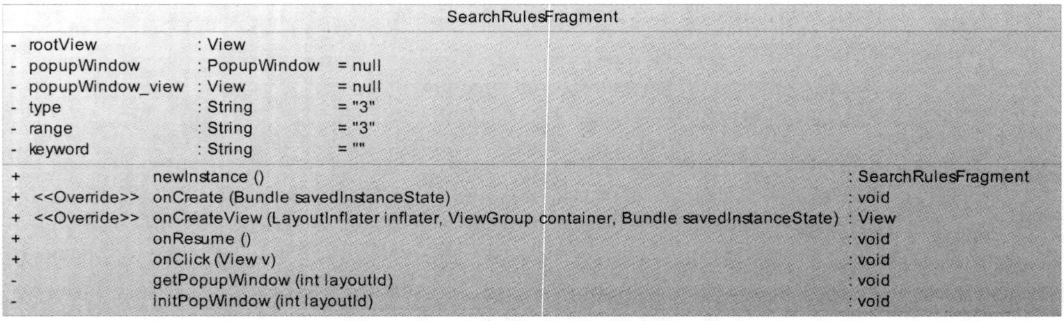

SearchRulesFragment		
- rootView	: View	
- popupWindow	: PopupWindow	= null
- popupWindow_view	: View	= null
- type	: String	= "3"
- range	: String	= "3"
- keyword	: String	= ""
+ newInstance ()		: SearchRulesFragment
+ <<Override>> onCreate (Bundle savedInstanceState)		: void
+ <<Override>> onCreateView (LayoutInflater inflater, ViewGroup container, Bundle savedInstanceState)		: View
+ onResume ()		: void
+ onClick (View v)		: void
- getPopupWindow (int layoutId)		: void
- initPopWindow (int layoutId)		: void

图 8.6　SearchRulesFragment 类图

加载页面布局与组件监听事件，代码如下：

```
rootView = inflater.inflate(R.layout.fragment_search_rules_main, container,
false);
```

```java
rootView.findViewById(R.id.image_top_layout_left).setOnClickListener(this);
rootView.findViewById(R.id.button_search).setOnClickListener(this);
//选择搜索商铺类型
rootView.findViewById(R.id.button_types_shops).setOnClickListener(this);
//选择搜索商铺地域
rootView.findViewById(R.id.button_dist_shops).setOnClickListener(this);
```

获得 PopupWindow，代码如下：

```java
/*
 * 获取 PopupWindow 实例
 */
private void getPopupWindow(int layoutId)
{
    if (null !=popupWindow)
    {
        popupWindow.dismiss();
        popupWindow =null;
    }
    initPopWindow(layoutId);
}

private void initPopWindow(int layoutId)
{
    popupWindow_view =View.inflate(getActivity(), layoutId, null);
    popupWindow =new PopupWindow(popupWindow_view, ViewGroup.LayoutParams.MATCH_PARENT, ViewGroup.LayoutParams.MATCH_PARENT, true);
    switch (layoutId)
    {
        case R.layout.popwindow_choice_types:
            popupWindow_view.findViewById(R.id.tv_type_one).setOnClickListener(this);
            popupWindow_view.findViewById(R.id.tv_type_two).setOnClickListener(this);
            popupWindow_view.findViewById(R.id.tv_type_all).setOnClickListener(this);
            popupWindow_view.findViewById(R.id.tv_type_all_cancel).setOnClickListener(this);
            break;
        case R.layout.popwindow_choice_distance:
            popupWindow_view.findViewById(R.id.tv_distance_one).setOnClickListener(this);
            popupWindow_view.findViewById(R.id.tv_distance_two).setOnClickListener(this);
```

```
            popupWindow_view.findViewById(R.id.tv_distance_all).setOnClickListener
(this);
            popupWindow_view.findViewById(R.id.tv_distance_cancel).
setOnClickListener(this);
            break;
    }
    ColorDrawable dw =new ColorDrawable(-00000);
    popupWindow.setBackgroundDrawable(dw);
    popupWindow.update();
}
```

传递数据到 SearchShopResultActivity,代码如下:

```
keyword = ((EditText) rootView.findViewById(R.id.edit_keywords_shops)).
getText()+"";
Intent intent =new Intent(getActivity(), SearchShopResultActivity.class);
intent.putExtra("type", type);
intent.putExtra("range", range);
intent.putExtra("keyword", keyword);
getActivity().startActivity(intent);
```

请求服务端并展现数据,此界面与全部商户页面大同小异。这里用的是 Activity,而全部商户用的是 Fragment。请对比两者的不同之处,这里不再详述。代码如下:

```
@Override
public void sendMessage(int opt)
{
    progressDialog.show();
    message =mHandler.obtainMessage();
    param =new JSONObject();
    try
    {
        switch (opt)
        {
            case OPT.SEARCH_SHOPS:
                param.put("act", "searchShop");
                param.put("keyword", keyword);
                param.put("type", type);
                if (null !=MeConfig.bdLocation)
                {
                param.put("latitude", MeConfig.bdLocation.getLatitude()+"");
                param.put("longitude", MeConfig.bdLocation.getLongitude()+"");
                }
                param.put("range", range);
                param.put("page", page);
                param.put("pageSize", pageSize);
```

```java
            break;
        }
    } catch (JSONException e)
    {
        e.printStackTrace();
    }
    Log.e(TAG, param.toString());
    super.sendMessage(opt);
}
```

sendMessage 请求数据。getData 获得数据后,更改界面。代码如下:

```java
@Override
public Object getData(int opt, Message msg)
{
    if (progressDialog !=null && progressDialog.isShowing())
    {
        progressDialog.dismiss();
    }
    switch (opt)
    {
        case OPT.SEARCH_SHOPS:
            ArrayList<HotsVendor>newsList = (ArrayList<HotsVendor>) msg.obj;
            if (newsList.size() <pageSize)
            {
                Toast.makeText(this, "数据已加载完毕!", Toast.LENGTH_LONG).show();
                isLastLoad =true;
            } else
            {
                page++;
            }
            isLoading =false;//获取完毕,解开锁定
            vendorList.addAll(newsList);
            shopListAdapter.notifyDataSetChanged();
            break;
    }
    return super.getData(opt, msg);
}
```

客户端与服务端的交互分为以下两步。
将数据封装到 JSON 对象中,代码如下:

```java
//商铺类型,"3"代表全部,"2"代表餐饮美食,"1"代表休闲娱乐
private String type ="3";
//距离范围"3"代表所有区域,"2"代表周围 1000 米范围,"1"代表周围 500 米
```

```java
private String range ="3";
//搜索关键词
private String keyword ="";

param.put("act", "searchShop");
param.put("keyword", keyword);
param.put("type", type);
if (null !=MeConfig.bdLocation)
{
    param.put("latitude", MeConfig.bdLocation.getLatitude() +"");
    param.put("longitude", MeConfig.bdLocation.getLongitude() +"");
}
param.put("range", range);
param.put("page", page);
param.put("pageSize", pageSize);
```

服务端返回 JSON String 解析放到 javabean 文件 HotsVendor.java 中,代码如下:

```java
package com.me.demo.bean;

import java.io.Serializable;
public class HotsVendor implements Serializable
{
    private static final long serialVersionUID =2282728744986789831L;
    public String shopId;
    public String shopAddress;
    public String shopName;
    public String shopTime;
    public String shopPhone;
    public String shopImage;
    public String shopLongitude;
    public String shopLatitude;
    public String shopType;
    public String shopDesc;
    public String shopSpend;
    public String distance;
}
```

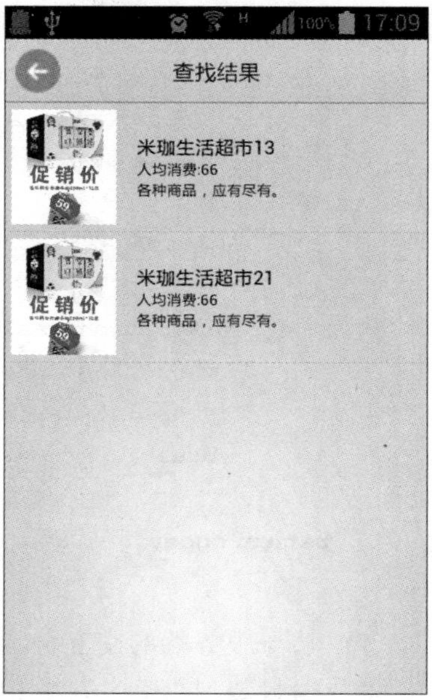

查找结果界面如图 8.7 所示。

商铺查找结果界面与全部商铺列表界面基本类似,主要由顶部 title 和内容展示部分的 ListView 组成。

图 8.7 查找结果界面

8.4 知识点回顾

本章主要知识点如下：
(1) 理解业务流程与实现方式。
(2) PopupWindow 的基本概念。

8.5 练　　习

(1) 实现对商户按类别搜索，并以列表形式分页展示搜索结果。
(2) 实现对商户按两个以上组合条件搜索，并以列表形式分页展示搜索结果。

第 9 章 与用户互动

9.1 让用户参与评价

9.1.1 用户发表评价的界面

用户发表评价的界面如图 9.1 所示。

图 9.1 用户发表评价的界面

如果需要定制该界面也可以自定义 RatingBar 组件,用不同的图片资源去实现不同的效果。

代码如下:

```
<RatingBar
        android:id="@+id/rating_bar_add_comment"
        style="@android:style/Widget.Holo.Light.RatingBar"
```

```
android:layout_width="wrap_content"
android:layout_height="wrap_content"
android:layout_marginLeft="10dp"
android:max="5"
android:rating="3.5"
android:stepSize="0.5" />
```

android：max＝"5"：设置 RatingBar 的最大值为 5 分。

android：rating＝"3.5"：设置当前显示的分数值为 3.5 分。

android：stepSize＝"0.5"：增加或减少为 0.5 的倍数。

当用户重新改变 RatingBar 的值后，需要提取 RatingBar 代表的值。

```
RatingBar pointRatingBar = (RatingBar) findViewById(R.id.rating_bar_add_comment);
appraisePoint =pointRatingBar.getRating() +"";
```

9.1.2 用户发表评价

1. 系统自带手机图片的裁剪

第 10 章会详细介绍如何利用系统裁剪图片，下面简单介绍该功能。

调用图片裁剪功能，如图 9.2 所示。图片裁剪成功后显示到界面上，如图 9.3 所示。

图 9.2　图片裁剪

图 9.3　图片裁剪成功后显示到界面上

调用系统裁剪工具 doCropPhoto 自定义方法，代码如下：

```java
protected void doCropPhoto()
{
    try
    {
        File uploadFileDir = new File(Environment.getExternalStorageDirectory(),
"meDemo" + File.separator + "upload");
        if (!uploadFileDir.exists())
        {
            uploadFileDir.mkdirs();
        }
        //Create a media file name
        String timeStamp = new SimpleDateFormat("yyyyMMdd_HHmmss", Locale.
getDefault()).format(new Date());
        picFile = new File(uploadFileDir, "img" + timeStamp + ".jpg");
        if (!picFile.exists())
        {
            picFile.createNewFile();
        }
        photoUri = Uri.fromFile(picFile);
        final Intent intent = getCropImageIntent();
        startActivityForResult(intent, activity_result_cropimage_with_data);
    } catch (Exception e)
    {
        e.printStackTrace();
    }
}

private Intent getCropImageIntent()
{
    Intent intent = new Intent(Intent.ACTION_PICK, photoUri);
    intent.setType("image/*");
    intent.putExtra("crop", "true");
    intent.putExtra("aspectX", 1);
    intent.putExtra("aspectY", 1);
    intent.putExtra("outputX", 320);
    intent.putExtra("outputY", 320);
    intent.putExtra("noFaceDetection", true);
    intent.putExtra("scale", true);
    intent.putExtra("return-data", false);
    intent.putExtra(MediaStore.EXTRA_OUTPUT, photoUri);
    intent.putExtra("outputFormat", Bitmap.CompressFormat.JPEG.toString());
    intent.setFlags(Intent.FLAG_ACTIVITY_SINGLE_TOP);
    return intent;
}
```

调用系统的裁剪工具后,系统裁剪完成会返回数据到 Activity。如果 Activity 要接收数据,则需要重写(override)onActivityResult 方法。

代码如下:

```java
@Override
protected void onActivityResult(int requestCode, int resultCode, Intent data)
{
    if (resultCode !=RESULT_OK || resultCode ==RESULT_CANCELED)
        return;
    switch (requestCode)
    {
        case activity_result_camara_with_data:      //拍照
            try
            {
                cropImageUriByTakePhoto();
            } catch (Exception e)
            {
                e.printStackTrace();
            }
            break;
        case activity_result_cropimage_with_data: //选择图片
            try
            {
                if (photoUri !=null)
                {
                    sendMessage(OPT.UPLODA_IMG);
                }
            } catch (Exception e)
            {
                e.printStackTrace();
            }
            break;
    }
    super.onActivityResult(requestCode, resultCode, data);
}
```

2. AddCommentActivity.java 解析

AddCommentActivity.java 类图如图 9.4 所示。

AddCommentActivity		
- popupWindow	: PopupWindow	= null
- popupWindow_view	: View	= null
- commentImage	: String	= null
- picFile	: File	
- photoUri	: Uri	
- filePath	: String	= null
- activity_result_camara_with_data	: int	= 1006
- activity_result_cropimage_with_data	: int	= 1007
- shopId	: String	
- appraiseContent	: String	
- appraisePoint	: String	= "3.5"
# <<Override>> onCreate (Bundle savedInstanceState)		: void
+ <<Override>> sendMessage (int opt)		: void
+ <<Override>> getData (int opt, Message msg)		: Object
+ onClick (View v)		: void
- getPopupWindow (int layoutId)		: void
- initPopWindow (int layoutId)		: void
# doTakePhoto ()		: void
# doCropPhoto ()		: void
- getCropImageIntent ()		: Intent
- cropImageUriByTakePhoto ()		: void
# onActivityResult (int requestCode, int resultCode, Intent data)		: void

图 9.4 AddCommentActivity 类图

代码如下：

package com.me.demo.activity;

import java.io.File;

import java.io.IOException;

import java.text.SimpleDateFormat;

import java.util.Date;

import java.util.Locale;

import org.json.JSONException;

import org.json.JSONObject;

import android.content.ActivityNotFoundException;

import android.content.Intent;

import android.graphics.Bitmap;

import android.graphics.drawable.ColorDrawable;

import android.net.Uri;

import android.os.Bundle;

import android.os.Environment;

import android.os.Message;

import android.provider.MediaStore;

import android.view.Gravity;

import android.view.View;

import android.view.View.OnClickListener;

import android.view.ViewGroup;

```java
import android.widget.Button;
import android.widget.EditText;
import android.widget.ImageView;
import android.widget.PopupWindow;
import android.widget.RatingBar;
import android.widget.Toast;

import com.me.demo.R;
import com.me.demo.application.MeApp;
import com.me.demo.util.BitmapUtil;
import com.me.demo.util.InfoTools;
import com.me.demo.util.OPT;
import com.me.demo.widget.WidgetTools;
public class AddCommentActivity extends BaseActivity implements OnClickListener
{
    private PopupWindow popupWindow =null;
    private View popupWindow_view =null;
    private String commentImage =null;
    private File picFile;
    private Uri photoUri;;
    private String filePath =null;
    private final int activity_result_camara_with_data =1006;
    private final int activity_result_cropimage_with_data =1007;
    private String shopId;
    private String appraiseContent;
    private String appraisePoint ="3.5";
    @Override
    protected void onCreate(Bundle savedInstanceState)
    {
        super.onCreate(savedInstanceState);
        Intent intent =this.getIntent();
        shopId =intent.getStringExtra("shopId");
        setContentView(R.layout.activity_add_comment_main);
        findViewById(R.id.text_top_layout_right).setOnClickListener(this);
        findViewById(R.id.image_top_layout_left).setOnClickListener(this);
        findViewById(R.id.button_choice_pic).setOnClickListener(this);
    }
    @Override
    public void sendMessage(int opt)
    {
        progressDialog.show();
        message =mHandler.obtainMessage();
        param =new JSONObject();
        try
```

```java
        {
            switch (opt)
            {
                case OPT.UPLODA_IMG:
                    filePath =BitmapUtil.savePic(picFile.getPath());
                    if (filePath ==null)
                    {
                        if (null !=progressDialog && progressDialog.isShowing())
                        {
                            progressDialog.dismiss();
                        }
                        return;
                    }
                    param.put("act", "postImage");
                    param.put("file", filePath);
                    break;
                case OPT.ADD_COMMENT:
                    param.put("act", "postAppraise");
                    param.put("appraiseContent", appraiseContent);
                    param.put("appraisePoint", appraisePoint);
                    param.put("appraiseImageUrl", commentImage);
                    param.put("userName", ((MeApp) getApplication()).userBean.username);
                    param.put("shopId", shopId);
                    break;
            }
        } catch (JSONException e)
        {
            e.printStackTrace();
        }
        super.sendMessage(opt);
    }
    @Override
    public Object getData(int opt, Message msg)
    {
        if (progressDialog !=null && progressDialog.isShowing())
        {
            progressDialog.dismiss();
        }
        switch (opt)
        {
            case OPT.UPLODA_IMG:
                commentImage =msg.obj.toString();
                ImageView imageView = (ImageView) findViewById(R.id.iv_upload_
```

```java
img);
                    mImageFetcher.loadImage(commentImage, imageView);
                break;
            case OPT.ADD_COMMENT:
                    Toast.makeText(this, msg.obj.toString(), Toast.LENGTH_LONG).show();
                    finish();
                break;
        }
        return super.getData(opt, msg);
    }
    @Override
    public void onClick(View v)
    {
        switch (v.getId())
        {
            case R.id.text_top_layout_right:
                //评分
                RatingBar pointRatingBar =(RatingBar) findViewById(R.id.rating_bar_add_comment);
                appraisePoint =pointRatingBar.getRating() +"";
                //内容
                EditText contentEdit = (EditText) findViewById(R.id.edit_comment_content);
                appraiseContent =contentEdit.getText() +"";
                if (appraiseContent.length() <10)
                {
                    WidgetTools.setTVError(contentEdit, this.getResources().getString(R.string.toast_comment_desc_tips), this);
                    return;
                }
                //图片地址
                if (commentImage ==null)
                {
                    WidgetTools.setTVError(((Button) findViewById(R.id.button_choice_pic)), this.getResources().getString(R.string.toast_shop_img_url_tips), this);
                    return;
                }
                sendMessage(OPT.ADD_COMMENT);
            break;
            case R.id.image_top_layout_left:
                finish();
            break;
```

```java
            case R.id.button_choice_pic:
                getPopupWindow(R.layout.popwindow_choice_picture);
                popupWindow.setInputMethodMode(PopupWindow.INPUT_METHOD_NOT_NEEDED);
                popupWindow.showAtLocation(v, Gravity.TOP, 0, 0);
                break;
            case R.id.mejust_from_photo:
                popupWindow.dismiss();
                if (!InfoTools.isSDCardAvailable())
                {
                    Toast.makeText(this, "未找到SDCard", Toast.LENGTH_LONG).show();
                    return;
                }
                doCropPhoto();
                break;
            case R.id.mejust_from_camera:
                popupWindow.dismiss();
                if (!InfoTools.isSDCardAvailable())
                {
                    Toast.makeText(this, "未找到SDCard", Toast.LENGTH_LONG).show();
                    return;
                }
                doTakePhoto();
                break;
            case R.id.mejust_from_cancle:
                popupWindow.dismiss();
                break;
            default:
                break;
        }
    }

    /*
     * 获取PopupWindow实例
     */
    private void getPopupWindow(int layoutId)
    {
        if (null !=popupWindow)
        {
            popupWindow.dismiss();
            popupWindow =null;
        }
        initPopWindow(layoutId);
```

```java
    }
    private void initPopWindow(int layoutId)
    {
        popupWindow_view =View.inflate(this, layoutId, null);
        popupWindow =new PopupWindow(popupWindow_view, ViewGroup.LayoutParams.
MATCH_PARENT, ViewGroup.LayoutParams.MATCH_PARENT, true);
        switch (layoutId)
        {
            case R.layout.popwindow_choice_picture:    popupWindow_view.
findViewById(R.id.mejust_from_photo).setOnClickListener(this);
popupWindow_view.findViewById(R.id.mejust_from_camera).setOnClickListener
(this);       popupWindow_view.findViewById(R.id.mejust_from_cancle).
setOnClickListener(this);
                break;
        }
        ColorDrawable dw =new ColorDrawable(-00000);
        popupWindow.setBackgroundDrawable(dw);
        popupWindow.update();
    }
    /**
     * 拍照获取图片
     */
    protected void doTakePhoto()
    {
        try
        {
            File uploadFileDir =new File(Environment.getExternalStorageDirectory(),
"meDemo" +File.separator +"upload" +File.separator);
            Intent cameraIntent =new Intent(MediaStore.ACTION_IMAGE_CAPTURE);
            if (!uploadFileDir.exists())
            {
                uploadFileDir.mkdirs();
            }
            //Create a media file name
            String timeStamp =new SimpleDateFormat("yyyyMMdd_HHmmss", Locale.
getDefault()).format(new Date());
            picFile =new File(uploadFileDir, "img" +timeStamp +".jpg");
            if (!picFile.exists())
            {
                picFile.createNewFile();
            }
            photoUri =Uri.fromFile(picFile);
            cameraIntent.putExtra(MediaStore.EXTRA_OUTPUT, photoUri);
            cameraIntent.setFlags(Intent.FLAG_ACTIVITY_SINGLE_TOP);
```

```java
            startActivityForResult(cameraIntent, activity_result_camara_with
_data);
        } catch (ActivityNotFoundException e)
        {
            e.printStackTrace();
        } catch (IOException e)
        {
            e.printStackTrace();
        }
    }
    protected void doCropPhoto()
    {
        try
        {
            File uploadFileDir = new File(Environment.getExternalStorageDirectory(),
"meDemo" +File.separator +"upload");
            if (!uploadFileDir.exists())
            {
                uploadFileDir.mkdirs();
            }

            //Create a media file name
            String timeStamp = new SimpleDateFormat("yyyyMMdd_HHmmss", Locale.
getDefault()).format(new Date());
            picFile = new File(uploadFileDir, "img" +timeStamp +".jpg");
            if (!picFile.exists())
            {
                picFile.createNewFile();
            }
            photoUri =Uri.fromFile(picFile);
            final Intent intent =getCropImageIntent();
            startActivityForResult(intent, activity_result_cropimage_with_
data);
        } catch (Exception e)
        {
            e.printStackTrace();
        }
    }

    /**
     * Constructs an intent for image cropping 调用图片剪辑程序
     */
    private Intent getCropImageIntent()
    {
```

```java
        Intent intent = new Intent(Intent.ACTION_PICK, photoUri);
        intent.setType("image/*");
        intent.putExtra("crop", "true");
        intent.putExtra("aspectX", 1);
        intent.putExtra("aspectY", 1);
        intent.putExtra("outputX", 320);
        intent.putExtra("outputY", 320);
        intent.putExtra("noFaceDetection", true);
        intent.putExtra("scale", true);
        intent.putExtra("return-data", false);
        intent.putExtra(MediaStore.EXTRA_OUTPUT, photoUri);
        intent.putExtra("outputFormat", Bitmap.CompressFormat.JPEG.toString());
        intent.setFlags(Intent.FLAG_ACTIVITY_SINGLE_TOP);
        return intent;
    }
    private void cropImageUriByTakePhoto()
    {
        Intent intent = new Intent("com.android.camera.action.CROP");
        intent.setDataAndType(photoUri, "image/*");
        intent.putExtra("crop", "true");
        intent.putExtra("aspectX", 1);
        intent.putExtra("aspectY", 1);
        intent.putExtra("outputX", 320);
        intent.putExtra("outputY", 320);
        intent.putExtra("scale", true);
        intent.putExtra(MediaStore.EXTRA_OUTPUT, photoUri);
        intent.putExtra("return-data", false);
        intent.putExtra("outputFormat", Bitmap.CompressFormat.JPEG.toString());
        intent.putExtra("noFaceDetection", true); //no face detection
        intent.setFlags(Intent.FLAG_ACTIVITY_SINGLE_TOP);
        startActivityForResult(intent, activity_result_cropimage_with_data);
    }
    @Override
    protected void onActivityResult(int requestCode, int resultCode, Intent data)
    {
        if (resultCode != RESULT_OK || resultCode == RESULT_CANCELED)
            return;
        switch (requestCode)
        {
            case activity_result_camara_with_data://拍照
                try
                {
                    cropImageUriByTakePhoto();
```

```
                } catch (Exception e)
                {
                    e.printStackTrace();
                }
                break;
            case activity_result_cropimage_with_data:
                try
                {
                    if (photoUri !=null)
                    {
                        sendMessage(OPT.UPLODA_IMG);
                    }
                } catch (Exception e)
                {
                    e.printStackTrace();
                }
                break;
        }
        super.onActivityResult(requestCode, resultCode, data);
    }
}
```

9.1.3 商户的评价列表展示

评价列表如图 9.5 所示，这里就不再讲解了，与商品列表整体差不多。

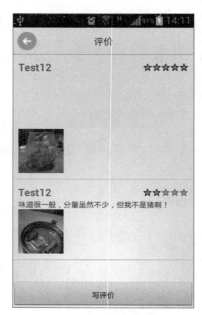

图 9.5 评价列表

9.2 让用户分享

9.2.1 什么是分享

在智能手机应用中已有"分享"功能。比如听到一首歌想要分享给好友时,就可以通过应用中的"分享"功能,分享到新浪微博、腾讯微博、QQ空间、微信等社交平台,呼叫好友围观、单击、转发、评论等。

这样就产生了一个社交传播闭环,更多用户看到分享的内容,进行分享评论的时候,会覆盖越来越多的粉丝。覆盖的粉丝会产生非常多的回流单击,回流单击又会产生社交单击,于是又会产生更多分享。

在应用中充分运用分享功能,优化分享转发、单击评论、用户回流的社会化传播闭环作用,所带来的价值潜力巨大。

9.2.2 让用户将内容分享到社交平台

从应用开发者的角度出发,通常他们自己开发这个分享功能,但由于每个分享平台需要分别配置,往往需要大量烦琐的工作。如果利用APP分享功能组件,直接粘贴一段代码就可以集成,可为开发者节省许多精力和时间。下面介绍iOS平台上非常热门的一款分享组件:ShareSDK。

ShareSDK是一种社会化分享组件,为iOS、Android、WP8的APP提供社会化功能,集成了一些常用的类库和接口,还有社会化统计分析管理后台,可缩短开发者的开发时间。

ShareSDK 支持 QQ、微信、新浪微博、腾讯微博、开心网、人人网、豆瓣、网易微博、搜狐微博、Facebook、Twitter、Google+等国内外40多家主流社交平台,可帮助开发者轻松实现社会化分享、登录、关注,及获得用户资料、好友列表等主流的社会化功能,强大的统计分析管理后台可以实时了解用户、信息流、回流率、传播效率等数据,有效地指导移动APP的日常运营与推广,同时也可为APP引入更多的社会化流量。

1. 系统分享

这种分享是一种很简单的分享模式,而且具有可选择性,对开发者以及用户是非常方便的一种方式。这种方式不需要集成额外的插件,一方面可以保证程序的安全性,另一方面也可减少程序的大小。

```
private void onShare()
{
    Intent intent = new Intent(Intent.ACTION_SEND);    //启动分享发送到属性
    intent.setType("text/plain");                       //分享发送到数据类型
    intent.putExtra(Intent.EXTRA_SUBJECT, "亲");         //分享的主题
    intent.putExtra(Intent.EXTRA_TEXT, "我的应用,非常棒。下载地址:www.xxx.com/
```

```
download.pak");        //分享的内容
    intent.setFlags(Intent.FLAG_ACTIVITY_SINGLE_TOP);
                                        //允许 Intent 启动新的 Activity
    startActivity(Intent.createChooser(intent, "分享"));
                                        //目标应用选择对话框的标题
}
```

2. 第三方分享

除了系统分享外,还有需要集成第三方插件的分享,这类分享普遍是比较大的平台。如果不采用这样的分享方式,将失去一个非常大的市场,比如微信朋友圈。如果不想失去这样一个圈子,就需要集成这个插件。

集成步骤如下:

(1) 登录 https://open.weixin.qq.com/网站,选择移动应用开发,微信接入流程如图 9.6 所示。

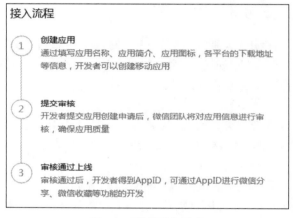

图 9.6　微信接入流程

(2) 在资源中心下载 Android 资源,如图 9.7 所示。

图 9.7　Android 资源下载

（3）引入开发包 libammsdk.jar，方法请参照前面章节介绍的如何引入第三方开发包。

（4）使用如下代码，完成微信开发的集成。

```java
package com.me.demo.util;

import java.net.URL;
import android.content.Context;
import android.graphics.Bitmap;
import android.graphics.BitmapFactory;
import android.os.Handler;
import android.os.Message;
import com.tencent.mm.sdk.openapi.IWXAPI;
import com.tencent.mm.sdk.openapi.SendMessageToWX;
import com.tencent.mm.sdk.openapi.WXAPIFactory;
import com.tencent.mm.sdk.openapi.WXImageObject;
import com.tencent.mm.sdk.openapi.WXMediaMessage;
import com.tencent.mm.sdk.openapi.WXTextObject;
import com.tencent.mm.sdk.platformtools.Util;
public class WeChatUtil
{
    private static final int MSG_FINISH_DOWNLOAD =1;
    private String appId ="此处填写你申请的应用 Id";
    private Context context =null;
    private IWXAPI api =null;
    private String url =null;
    private String text =null;
    public WeChatUtil(Context context, String text)
    {
        this.context =context;
        this.text =text;
    }
    private Handler handler =new Handler()
    {
        @Override
        public void handleMessage(Message msg)
        {
            switch (msg.what)
            {
                case MSG_FINISH_DOWNLOAD:
                    try
                    {
                        Bitmap bmp = (Bitmap) msg.obj;
                        WXImageObject imageObject =new WXImageObject();
```

```java
                            imageObject.imageUrl = url;

                            WXMediaMessage wxmsg = new WXMediaMessage();
                            wxmsg.mediaObject = imageObject;
                            Bitmap thumbBmp = Bitmap.createScaledBitmap(bmp, 150, 150,
    true);
                            if (bmp != null && (!bmp.isRecycled()))
                            {
                                bmp.recycle();
                            }
                            wxmsg.thumbData = Util.bmpToByteArray(thumbBmp, true);
                            wxmsg.description = text;
                            wxmsg.title = text;
                            SendMessageToWX.Req req = new SendMessageToWX.Req();
                            req.transaction = buildTransaction("img");
                            req.message = wxmsg;
                            req.scene = SendMessageToWX.Req.WXSceneTimeline;
                            if (!api.sendReq(req))
                            {
                                throw new Exception("发送请求失败!");
                            } else
                            {
                                System.out.println("分享图片成功,地址:" + url);
                            }
                        } catch (Exception e)
                        {
                            e.printStackTrace();
                        } finally
                        {
                            api.unregisterApp();
                        }
                        break;
                    default:
                        super.handleMessage(msg);
                        break;
                }
            }
        };
        class WorkingThread extends Thread
        {
            private String url = null;
            public WorkingThread(String url)
            {
                this.url = url;
```

```java
        }
        @Override
        public void run()
        {
            try
            {
                Bitmap bmp =BitmapFactory.decodeStream(new URL(url).openStream());
                if (null ==bmp)
                {
                    throw new Exception("图片解码失败!");
                }
                //通知主线程完成更新
                Message msg =WeChatUtil.this.handler.obtainMessage(MSG_FINISH_DOWNLOAD);
                msg.obj =bmp;
                WeChatUtil.this.handler.sendMessage(msg);
            } catch (Exception e)
            {
                e.printStackTrace();
            }
        }
    }
    /**
     * 发送网络图片
     */
    public void sendHttpImage(String url)
    {
        this.url =url;
        api =WXAPIFactory.createWXAPI(context, appId);
        try
        {
            boolean wxInstall =api.isWXAppInstalled();
            if (!wxInstall)
            {
                throw new Exception("微信未安装!");
            }

            if (!api.registerApp(appId))
            {
                throw new Exception("微信 API 注册失败!");
            }
            WorkingThread t =new WorkingThread(url);
            t.start();
        } catch (Exception e)
```

```java
            {
                e.printStackTrace();
            }
        }
    private String buildTransaction(final String type)
    {
        return (type == null) ? String.valueOf(System.currentTimeMillis()) :
type + System.currentTimeMillis();
    }

    /**
     * 发送文字
     *
     * @param text
     */
    public void sendText()
    {
        api = WXAPIFactory.createWXAPI(context, appId);
        try
        {
            boolean wxInstall = api.isWXAppInstalled();
            if (!wxInstall)
            {
                throw new Exception("微信未安装!");
            }

            if (!api.registerApp(appId))
            {
                throw new Exception("微信 API 注册失败!");
            }
            //初始化一个 WXTextObject 对象
            WXTextObject textObj = new WXTextObject();
            textObj.text = text;
            //用 WXTextObject 对象初始化一个 WXMediaMessage 对象
            WXMediaMessage msg = new WXMediaMessage();
            msg.mediaObject = textObj;
            //发送文本类型的消息时,title 字段不起作用
            //msg.title = "Will be ignored";
            msg.description = text;
            //构造一个 Req
            SendMessageToWX.Req req = new SendMessageToWX.Req();
            req.transaction = buildTransaction("text"); //transaction 字段用于
唯一标识一个请求
            req.message = msg;
```

```
            req.scene =SendMessageToWX.Req.WXSceneTimeline;
            //调用 API 接口发送数据到微信
            if (!api.sendReq(req))
                System.out.println("短消息发送失败!");
            api.unregisterApp();
        } catch (Exception e)
        {
            e.printStackTrace();
        }
    }
}
```

9.3 给用户推送消息

9.3.1 推送的几种常见解决方案

推送的几种常见解决方案如下:

(1) 轮询(Pull)方式:应用程序应当阶段性地与服务器进行连接并查询是否有新的消息到达,必须自己实现与服务器之间的通信,例如消息排队等;还要考虑轮询的频率。如果太慢,可能导致某些消息的延迟;如果太快,则会大量消耗网络带宽和电池。

(2) SMS(Push)方式:在 Android 平台上,可以通过拦截 SMS 消息并且解析消息内容来了解服务器的意图,并获取其显示内容进行处理。这个方案的好处是,可以实现完全的实时操作。问题是这个方案的成本相对比较高,需要向移动公司缴纳相应的费用。目前很难找到免费的短消息发送网关来实现这种方案。

(3) 持久连接(Push)方式:这个方案可以解决由轮询带来的性能问题,但还是会消耗手机的电池。iOS 平台的推送服务之所以很好,是因为每一台手机仅仅保持一个与服务器之间的连接,事实上 C2DM 也是如此。这个方案存在很多不足之处,很难在手机上实现一个可靠的服务,目前也无法与 iOS 平台的推送功能相比。

Android 操作系统允许在低内存情况下终止系统服务,所以我们的推送通知服务很有可能被操作系统终止了。轮询(Pull)方式和 SMS(Push)方式存在明显的不足。持久连接(Push)方案也有不足,不过可以通过良好的设计来弥补,以便让该方案可以有效地工作。毕竟 GMail、GTalk 以及 GoogleVoice 都可以实现实时更新。

推送方案有很多,最重要的处理工作不是在 Android 客户端,而是在服务端的功能开发上。这里不详述整个推送服务的实现,而是讲解现在已经很成熟的第三方推送平台,它对于实现自己的业务需求非常方便。

9.3.2 常用的推送平台

这里主要介绍两种推送平台:极光推送和百度云推送。

1. 极光推送

集成步骤如下：

（1）登录 https://www.jpush.cn/，注册账号，申请应用 ID，下载 JPush Android SDK。

（2）导入 SDK 开发包到自己的应用程序项目。解压缩 jpush-sdk_v1.x.y.zip 集成压缩包，复制 libs/jpush-sdk-release1.x.y.jar 到工程 libs/目录中，复制 libs/armeabi/libjpush1xy.so 到工程 libs/armeabi 目录中。如果项目中有 libs/armeabi-v7a 这个目录，请把 armeabi 的 so 文件也复制一份到这个目录中。

（3）配置 AndroidManifest.xml。

（4）根据 SDK 压缩包里的 AndroidManifest.xml 样例文件，配置应用程序项目的 AndroidManifest.xml。

下面介绍配置文件的方法与步骤。

复制备注为"Required"的部分，将备注为替换包名的部分替换为当前应用程序的包名，将 AppKey 替换为在 Portal 上注册该应用的 Key，例如（9fed5bcb7b9b87413678c407），配置权限。

代码如下：

```xml
<?xml version="1.0" encoding="utf-8"?>
<manifest xmlns:android="http://schemas.android.com/apk/res/android"
    package="Your Package"
    android:versionCode="100"
    android:versionName="1.0.0" >
    <!--Required -->
    <permission
        android:name="Your Package.permission.JPUSH_MESSAGE"
        android:protectionLevel="signature" />
    <!--Required -->
    <uses-permission android:name="You Package.permission.JPUSH_MESSAGE" />
    <uses-permission android:name="android.permission.RECEIVE_USER_PRESENT" />
    <uses-permission android:name="android.permission.INTERNET" />
    <uses-permission android:name="android.permission.WAKE_LOCK" />
    <uses-permission android:name="android.permission.READ_PHONE_STATE" />
    <uses-permission android:name="android.permission.WRITE_EXTERNAL_STORAGE" />
    <uses-permission android:name="android.permission.READ_EXTERNAL_STORAGE" />
    <uses-permission android:name="android.permission.VIBRATE" />
    <uses-permission android:name="android.permission.MOUNT_UNMOUNT_FILESYSTEMS" />
    <uses-permission android:name="android.permission.ACCESS_NETWORK_STATE" />
    <uses-permission android:name="android.permission.SYSTEM_ALERT_WINDOW" />
    <uses-permission android:name="android.permission.WRITE_SETTINGS" />
```

```xml
<!--since 1.6.0 -->
    <!--Optional. Required for location feature -->
    <uses-permission android:name="android.permission.ACCESS_COARSE_LOCATION" />
    <uses-permission android:name="android.permission.ACCESS_COARSE_UPDATES" />
    <uses-permission android:name="android.permission.ACCESS_WIFI_STATE" />
    <uses-permission android:name="android.permission.CHANGE_WIFI_STATE" />
    <uses-permission android:name="android.permission.ACCESS_FINE_LOCATION" />
    <uses-permission android:name="android.permission.ACCESS_LOCATION_EXTRA_COMMANDS" />
    <uses-permission android:name="android.permission.CHANGE_NETWORK_STATE" />
    <application
        android:name="Your Application"
        android:icon="@drawable/ic_launcher"
        android:label="@string/app_name" >
        <!--Required -->
        <service
            android:name="cn.jpush.android.service.PushService"
            android:enabled="true"
            android:exported="false" >
            <intent-filter>
                <action android:name="cn.jpush.android.intent.REGISTER" />
                <action android:name="cn.jpush.android.intent.REPORT" />
                <action android:name="cn.jpush.android.intent.PushService" />
                <action android:name="cn.jpush.android.intent.PUSH_TIME" />
            </intent-filter>
        </service>

        <!--Required -->
        <receiver
            android:name="cn.jpush.android.service.PushReceiver"
            android:enabled="true" >
            <intent-filter android:priority="1000" >
<!--since 1.3.5 -->
                <action android:name="cn.jpush.android.intent.NOTIFICATION_RECEIVED_PROXY" />
<!--since 1.3.5 -->
                <category android:name="Your Package" />
<!--since 1.3.5 -->
            </intent-filter>
<!--since 1.3.5 -->
            <intent-filter>
                <action android:name="android.intent.action.USER_PRESENT" />
```

```xml
            <action android:name="android.net.conn.CONNECTIVITY_CHANGE" />
        </intent-filter>
        <intent-filter>
            <action android:name="android.intent.action.PACKAGE_ADDED" />
            <action android:name="android.intent.action.PACKAGE_REMOVED" />
            <data android:scheme="package" />
        </intent-filter>
    </receiver>
    <!--Required SDK 核心功能 -->
    <activity
        android:name="cn.jpush.android.ui.PushActivity"
        android:configChanges="orientation|keyboardHidden"
        android:theme="@android:style/Theme.Translucent.NoTitleBar" >
        <intent-filter>
            <action android:name="cn.jpush.android.ui.PushActivity" />
            <category android:name="android.intent.category.DEFAULT" />
            <category android:name="Your Package" />
        </intent-filter>
    </activity>
    <!--Required SDK 核心功能 -->
    <service
        android:name="cn.jpush.android.service.DownloadService"
        android:enabled="true"
        android:exported="false" >
    </service>
    <!--Required SDK 核心功能 -->
    <receiver android:name="cn.jpush.android.service.AlarmReceiver" />

    <!--Required. For publish channel feature -->
    <!--JPUSH_CHANNEL 是为了方便开发者统计 APK 分发渠道 -->
    <!--例如：-->
    <!--发到 Google Play 的 APK 可以设置为 google-play; -->
    <!--发到其他市场的 APK 可以设置为 xxx-market -->
    <!--目前这个渠道统计功能的报表还未开放 -->
    <meta-data
        android:name="JPUSH_CHANNEL"
        android:value="developer-default" />
    <!--Required. AppKey copied from Portal -->
    <meta-data
        android:name="JPUSH_APPKEY"
        android:value="Your AppKey" />
</application>
</manifest>
```

添加代码，JPush SDK 提供的 API 接口主要都集中在 cn.jpush.android.api.

JPushInterface 类中。init 初始化 SDK：

public static void init(Context context)

setDebugMode 设置调试模式，代码如下：

```
//You can enable debug mode in developing state. You should close debug mode when release.
```
public static void setDebugMode(**boolean** debugEnalbed);

2. 百度云推送

集成步骤如下：
(1) 登录 http://developer.baidu.com/cloud/push。
(2) 进入移动应用管理中心。
(3) 创建应用获取 API key。
(4) 下载 Android 客户端 SDK。
(5) 将 SDK 的 libs 中的.jar 和.so 文件复制到新建的项目中，如图 9.8 所示。

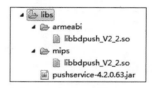

图 9.8　libs 目录

(6) 自定义 MeApp 继承 com.baidu.frontia.FrontiaApplication，并配置到 manifest 中。

```
<uses-permission android:name="android.permission.INTERNET" />
    <uses-permission android:name="android.permission.READ_PHONE_STATE" />
    <uses-permission android:name="android.permission.ACCESS_NETWORK_STATE" />
    <uses-permission android:name="android.permission.RECEIVE_BOOT_COMPLETED" />
    <uses-permission android:name="android.permission.WRITE_SETTINGS" />
    <uses-permission android:name="android.permission.VIBRATE" />
    <uses-permission android:name="android.permission.WRITE_EXTERNAL_
STORAGE" />
    <uses-permission android:name="android.permission.DISABLE_KEYGUARD" />
    <uses-permission android:name="android.permission.ACCESS_COARSE_
LOCATION" />
    <uses-permission android:name="android.permission.ACCESS_WIFI_STATE" />
```

(7) 注册两个 Receiver 和一个 Service。

```
<receiver android:name="com.baidu.push.example.MyPushMessageReceiver" >
    <intent-filter>
<!--接收 push 消息 -->
    <action android:name="com.baidu.android.pushservice.action.MESSAGE" />
            <!--接收 bind,unbind,fetch,delete 等反馈消息 -->
    <action android:name="com.baidu.android.pushservice.action.RECEIVE" />
    <action android:name="com.baidu.android.pushservice.action.notification.
CLICK" />
    </intent-filter>
</receiver>
```

```xml
<!--push 必需的 receiver 和 service 声明 -->
<receiver 
        android:name="com.baidu.android.pushservice.PushServiceReceiver"
    android:process=":bdservice_v1" >
    <intent-filter>
    <action android:name="android.intent.action.BOOT_COMPLETED" />
    <action android:name="android.net.conn.CONNECTIVITY_CHANGE" />
     <action android:name="com.baidu.android.pushservice.action.notification.SHOW" />
    <action android:name="com.baidu.android.pushservice.action.media.CLICK" />
    </intent-filter>
</receiver>
    <receiver 
        android:name="com.baidu.android.pushservice.RegistrationReceiver"
    android:process=":bdservice_v1" >
    <intent-filter>
    <action android:name="com.baidu.android.pushservice.action.METHOD" />
    <action android:name="com.baidu.android.pushservice.action.BIND_SYNC" />
    </intent-filter>
    <intent-filter>
    <action android:name="android.intent.action.PACKAGE_REMOVED" />
        <data android:scheme="package" />
    </intent-filter>
</receiver>
<service
   android:name="com.baidu.android.pushservice.PushService"
   android:exported="true"
   android:process=":bdservice_v1" >
   <intent-filter>
        <action android:name="com.baidu.android.pushservice.action.PUSH_SERVICE" />
   </intent-filter>
</service>
```

（8）官方 demo 源代码。

```java
package com.baidu.push.example;

import java.util.List;
import android.app.Activity;
import android.app.AlertDialog;
import android.app.Notification;
import android.content.DialogInterface;
import android.content.Intent;
```

```java
import android.content.res.Resources;
import android.os.Bundle;
import android.util.Log;
import android.view.View;
import android.webkit.CookieManager;
import android.webkit.CookieSyncManager;
import android.widget.Button;
import android.widget.EditText;
import android.widget.LinearLayout;
import android.widget.RelativeLayout;
import android.widget.ScrollView;
import android.widget.TextView;

import com.baidu.android.pushservice.CustomPushNotificationBuilder;
import com.baidu.android.pushservice.PushConstants;
import com.baidu.android.pushservice.PushManager;

/*
 * 云推送 Demo 主 Activity
 * 代码中,注释以 Push 标注开头的,表示接下来的代码块是 Push 接口调用示例
 */
public class PushDemoActivity extends Activity implements View.OnClickListener
{
    private static final String TAG = PushDemoActivity.class.getSimpleName();
    RelativeLayout mainLayout = null;
    int akBtnId = 0;
    int initBtnId = 0;
    int richBtnId = 0;
    int setTagBtnId = 0;
    int delTagBtnId = 0;
    int clearLogBtnId = 0;
    Button initButton = null;
    Button initWithApiKey = null;
    Button displayRichMedia = null;
    Button setTags = null;
    Button delTags = null;
    Button clearLog = null;
    TextView logText = null;
    ScrollView scrollView = null;
    public static int initialCnt = 0;
    private boolean isLogin = false;

    @Override
    public void onCreate(Bundle savedInstanceState)
```

```java
    {
        super.onCreate(savedInstanceState);

        Utils.logStringCache =Utils.getLogText(getApplicationContext());

        Resources resource =this.getResources();
        String pkgName =this.getPackageName();

        setContentView(resource.getIdentifier("main", "layout", pkgName));
        akBtnId =resource.getIdentifier("btn_initAK", "id", pkgName);
        initBtnId =resource.getIdentifier("btn_init", "id", pkgName);
        richBtnId =resource.getIdentifier("btn_rich", "id", pkgName);
        setTagBtnId =resource.getIdentifier("btn_setTags", "id", pkgName);
        delTagBtnId =resource.getIdentifier("btn_delTags", "id", pkgName);
        clearLogBtnId =resource.getIdentifier("btn_clear_log", "id", pkgName);

        initWithApiKey = (Button) findViewById(akBtnId);
        initButton = (Button) findViewById(initBtnId);
        displayRichMedia = (Button) findViewById(richBtnId);
        setTags = (Button) findViewById(setTagBtnId);
        delTags = (Button) findViewById(delTagBtnId);
        clearLog = (Button) findViewById(clearLogBtnId);

        logText = (TextView) findViewById(resource.getIdentifier("text_log", "id", pkgName));
        scrollView = (ScrollView) findViewById(resource.getIdentifier("stroll_text", "id", pkgName));

        initWithApiKey.setOnClickListener(this);
        initButton.setOnClickListener(this);
        setTags.setOnClickListener(this);
        delTags.setOnClickListener(this);
        displayRichMedia.setOnClickListener(this);
        clearLog.setOnClickListener(this);

        //Push: 以 apikey 的方式登录,一般放在主 Activity 的 onCreate 中
        //这里把 apikey 存放于 manifest 文件中,只是一种存放方式
        //可以用自定义常量等其他方式实现,来替换参数中的 Utils.getMetaValue(PushDemoActivity.this, "api_key")
        PushManager.startWork(getApplicationContext(), PushConstants.LOGIN_TYPE_API_KEY, Utils.getMetaValue(PushDemoActivity.this, "api_key"));
        //Push: 如果想基于地理位置推送,可以打开支持地理位置的推送的开关
        //PushManager.enableLbs(getApplicationContext());
```

```java
        //Push:设置自定义的通知样式,具体 API 介绍见用户手册,如果想使用系统默认的可
以不加这段代码
        //请在通知推送界面中,选择"高级设置"→"通知栏样式"→"自定义样式",并且填写
值:1
        //与下面代码 PushManager.setNotificationBuilder(this, 1, cBuilder)中的
第二个参数对应
            CustomPushNotificationBuilder cBuilder = new CustomPushNotificationBuilder
(getApplicationContext(), resource.getIdentifier("notification_custom_
builder", "layout", pkgName), resource.getIdentifier(
            "notification_icon", "id", pkgName), resource.getIdentifier("
notification_title", "id", pkgName), resource.getIdentifier("notification_
text", "id", pkgName));
            cBuilder.setNotificationFlags(Notification.FLAG_AUTO_CANCEL);
            cBuilder.setNotificationDefaults(Notification.DEFAULT_SOUND | Notification.
DEFAULT_VIBRATE);
            cBuilder.setStatusbarIcon(this.getApplicationInfo().icon);
            cBuilder.setLayoutDrawable(resource.getIdentifier("simple_notification_
icon", "drawable", pkgName));
            PushManager.setNotificationBuilder(this, 1, cBuilder);
    }

    @Override
    public void onClick(View v)
    {
        if (v.getId() ==akBtnId)
        {
            initWithApiKey();
        } else if (v.getId() ==initBtnId)
        {
            initWithBaiduAccount();
        } else if (v.getId() ==richBtnId)
        {
            openRichMediaList();
        } else if (v.getId() ==setTagBtnId)
        {
            setTags();
        } else if (v.getId() ==delTagBtnId)
        {
            deleteTags();
        } else if (v.getId() ==clearLogBtnId)
        {
            Utils.logStringCache = "";
            Utils.setLogText(getApplicationContext(), Utils.logStringCache);
            updateDisplay();
```

```java
            }
        }

        //打开富媒体列表界面
        private void openRichMediaList()
        {
            //Push：打开富媒体消息列表
            Intent sendIntent = new Intent();
            sendIntent.setClassName(getBaseContext(), "com.baidu.android.pushservice.richmedia.MediaListActivity");
            sendIntent.addFlags(Intent.FLAG_ACTIVITY_NEW_TASK);
            PushDemoActivity.this.startActivity(sendIntent);
        }

        //删除tag操作
        private void deleteTags()
        {
            LinearLayout layout = new LinearLayout(PushDemoActivity.this);
            layout.setOrientation(LinearLayout.VERTICAL);

            final EditText textviewGid = new EditText(PushDemoActivity.this);
            textviewGid.setHint("请输入多个标签,以英文逗号隔开");
            layout.addView(textviewGid);

            AlertDialog.Builder builder = new AlertDialog.Builder(PushDemoActivity.this);
            builder.setView(layout);
            builder.setPositiveButton("删除标签", new DialogInterface.OnClickListener()
            {
                public void onClick(DialogInterface dialog, int which)
                {
                    //Push：删除tag调用方式
                    List<String> tags = Utils.getTagsList(textviewGid.getText().toString());
                    PushManager.delTags(getApplicationContext(), tags);
                }
            });
            builder.show();
        }

        //设置标签,以英文逗号隔开
        private void setTags()
        {
            LinearLayout layout = new LinearLayout(PushDemoActivity.this);
```

```java
        layout.setOrientation(LinearLayout.VERTICAL);

        final EditText textviewGid = new EditText(PushDemoActivity.this);
        textviewGid.setHint("请输入多个标签,以英文逗号隔开");
        layout.addView(textviewGid);

        AlertDialog.Builder builder = new AlertDialog.Builder(PushDemoActivity.this);
        builder.setView(layout);
        builder.setPositiveButton("设置标签", new DialogInterface.OnClickListener()
        {
            public void onClick(DialogInterface dialog, int which)
            {
                //Push: 设置 tag 调用方式
                List<String> tags = Utils.getTagsList(textviewGid.getText().toString());
                PushManager.setTags(getApplicationContext(), tags);
            }
        });
        builder.show();
    }

    //以 api_key 的方式绑定
    private void initWithApiKey()
    {
        //Push: 无账号初始化,用 api_key 绑定
        PushManager.startWork(getApplicationContext(), PushConstants.LOGIN_TYPE_API_KEY, Utils.getMetaValue(PushDemoActivity.this, "api_key"));
    }

    //以百度账号登录,获取 access token 来绑定
    private void initWithBaiduAccount()
    {
        if (isLogin)
        {
            //已登录则清除 Cookie, access token, 设置"登录"按钮
            CookieSyncManager.createInstance(getApplicationContext());
            CookieManager.getInstance().removeAllCookie();
            CookieSyncManager.getInstance().sync();

            isLogin = false;
            initButton.setText("登录百度账号初始化 Channel");
        }
        //跳转到百度账号登录的 activity
```

```java
        Intent intent = new Intent(PushDemoActivity.this, LoginActivity.class);
        startActivity(intent);
    }


    @Override
    public void onResume()
    {
        super.onResume();
        Log.d(TAG, "onResume");
        updateDisplay();
    }


    @Override
    protected void onNewIntent(Intent intent)
    {
        String action = intent.getAction();
        if (Utils.ACTION_LOGIN.equals(action))
        {
            //Push：百度账号初始化，用 access token 绑定
            String accessToken = intent.getStringExtra(Utils.EXTRA_ACCESS_TOKEN);
            PushManager.startWork(getApplicationContext(), PushConstants.LOGIN_TYPE_ACCESS_TOKEN, accessToken);
            isLogin = true;
            initButton.setText("更换百度账号");
        }
        updateDisplay();
    }


    @Override
    public void onDestroy()
    {
        Utils.setLogText(getApplicationContext(), Utils.logStringCache);
        super.onDestroy();
    }


    //更新界面显示内容
    private void updateDisplay()
    {
        Log.d(TAG, "updateDisplay, logText:" + logText + " cache: " + Utils.logStringCache);
        if (logText != null)
        {
            logText.setText(Utils.logStringCache);
        }
```

```
        if (scrollView !=null)
        {
            scrollView.fullScroll(ScrollView.FOCUS_DOWN);
        }
    }
}
```

9.4 知识点回顾

本章主要知识点如下：
（1）RatingBar 的理解与使用。
（2）系统自带图片的裁剪与使用。
（3）分享的基本概念。
（4）推送的几种方式。

9.5 练　　习

（1）实现注册用户对商户（商品）的评分。
（2）实现注册用户对商户（商品）发表 100 字以内的评价。

第 10 章 添加商户信息

10.1 添加商户信息总体设计

添加商户时序如图 10.1 所示。

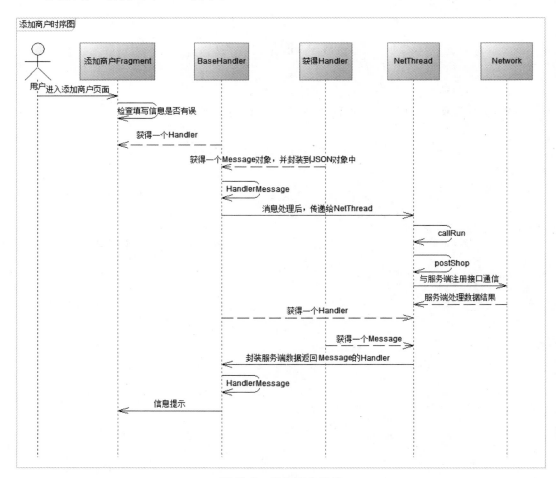

图 10.1 添加商户时序

添加商户流程如图 10.2 所示。

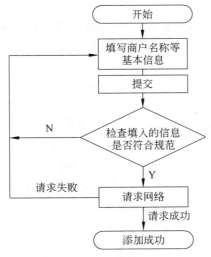

图 10.2 添加商户流程

10.2 商户数据库准备

详细情况请参见 6.2 节。

10.3 Intent 详解

1. 什么是 Intent

Intent 是 Android 的核心和灵魂,是各组件之间的桥梁。Android 应用的四大组件分别为 Activity、Service、BroadcastReceiver、ContentProvider。四种组件是相互独立的,它们之间可以互相调用、协调工作,最终组成一个真正的 Android 应用。

通过 Intent,应用程序可以向 Android 表达某种请求或者意愿,Android 会根据意愿的内容选择适当的组件来完成请求。比如,有一个 Activity 希望打开网页浏览器查看某一网页的内容,那么这个 Activity 只需要发出 WEB_SEARCH_ACTION 给 Android,Android 就会根据 Intent 的请求内容,查询各组件注册时声明的 IntentFilter,找到网页浏览器的 Activity 来浏览网页。

Android 的三个基本组件 Activity、Service 和 Broadcast Receiver 都是通过 Intent 机制激活的,不同类型的组件有不同的传递 Intent 方式。Intent 一旦发出,Android 都会准确找到与其相匹配的一个或多个 Activity、Service 或 BroadcastReceiver 给出响应,所以不同类型的 Intent 消息不会出现重叠,即 Broadcast 的 Intent 消息只会发送给 BroadcastReceiver,而决不会发送给 Activity 或者 Service。由 startActivity()传递的消

息只会发送给 Activity，由 startService()传递的 Intent 也只会发送给 Service。

1）Activity Intent

要激活一个新的 Activity，或者让一个已有的 Activity 做新的操作，可以通过调用 Context.startActivity()或者 Activity.startActivityForResult()方法实现。

2）Service Intent

要启动一个新的 Service，或者向一个已有的 Service 传递新的指令，可调用 Context.startService()方法或者调用 Context.bindService()方法将调用此方法的上下文对象与 Service 绑定。

3）Broadcast Intent

Context.sendBroadcast()、Context.sendOrderBroadcast()、Context.sendStickBroadcast()这三个方法可以发送 Broadcast Intent。发送之后，所有已注册并且拥有与之匹配 IntentFilter 的 BroadcastReceiver 就会被激活。

2. Intent 的构成

要在不同的 activity 之间传递数据，就要在 intent 中包含相应的内容，一般来说数据中主要应该包括下述内容。

1）Action

Action 用来指明要实施的动作是什么，比如说 ACTION_VIEW、ACTION_EDIT 等，具体内容可以查阅 android SDK→reference 中的 Android.content.intent 类，里面的 constant 中定义了所有的 Action。

一些常用的 Action 如下：

- ACTION_CALL activity：启动一个电话。
- ACTION_EDIT activity：显示用户编辑的数据。
- ACTION_MAIN activity：作为 Task 中第一个 Activity 启动。
- ACTION_SYNC activity：同步手机与数据服务器上的数据。
- ACTION_BATTERY_LOW broadcast receiver：电池电量过低警告。
- ACTION_HEADSET_PLUG broadcast receiver：插拔耳机警告。
- ACTION_SCREEN_ON broadcast receiver：屏幕变亮警告。
- ACTION_TIMEZONE_CHANGED broadcast receiver：改变时区警告。

2）Data

Data 是具体的数据，一般由一个 URI 变量来表示。

3）Category

Category 是一个字符串，包含了关于处理该 Intent 的组件的种类的信息。一个 Intent 对象可以有任意个 Category。Intent 类定义了许多 Category 常数。

- addCategory()：为一个 Intent 对象增加一个 Category。
- removeCategory：删除一个 Category。
- getCategories()：获取 Intent 所有的 Category。

4）Type

显式指定 Intent 的数据类型 MIME(Multipurpose Internet Mail Extensions，多用途互联

网邮件扩展)。比如,一个组件可以显示图片数据而不能播放声音文件。很多情况下,数据类型可在 URI 中找到,比如 content:开头的 URI,表明数据由设备上的 content provider 提供。通过设置这个属性,可以强制采用显式指定的类型而不再进行推导。

MIME 类型有两种形式,分别介绍如下。

(1) 单个记录的格式:vnd. android. cursor. item/vnd. yourcompanyname. contenttype。例如,content://com. example. transportationprovider/trains/122(一辆列车信息的 URI)的 MIME 类型是 vnd. android. cursor. item/vnd. example. rail。

(2) 多个记录的格式:vnd. android. cursor. dir/vnd. yourcompanyname. contenttype。例如,content://com. example. transportationprovider/trains(所有列车信息的 RUI)的 MIME 类型是 vnd. android. cursor. dir/vnd. example. rail。

5) Component

Component 指定 Intent 的目标组件的类名称。通常 Android 会根据 Intent 中包含的其他属性的信息,比如 Action、Data、Type、Category 进行查找,最终找到一个与之匹配的目标组件。如果 Component 这个属性已指定了组件,则直接使用它指定的组件,而不再执行上述查找过程。指定了这个属性以后,Intent 的其他所有属性都是可选的。例如:

```
Intent it = new Intent (Activity.Main.this, Activity2.class); startActivity
(it);
startActivity(it);
```

6) Extras

Extras 是附加信息。例如 ACTION_TIMEZONE_CHANGED 的 intent 有一个 "time-zone"附加信息来指明新的时区,而 ACTION_HEADSET_PLUG 有一个"state"附加信息来指示耳机是被插入还是被拔出。Intent 对象有一系列 put...()和 set...()方法来设定和获取附加信息。这些方法和 Bundle 对象类似。事实上附加信息可以使用 putExtras()和 getExtras()作为 Bundle 来读和写。例如:用 Bundle 传递数据:

```
Intent it =new Intent(Activity.Main.this, Activity2.class); Bundle bundle=new
Bundle(); bundle.putString("name", "This is from MainActivity!"); it.putExtras
(bundle); startActivity(it); //获得数据 Bundle bundle=getIntent().getExtras();
String name=bundle.getString("name");
```

3. Intent 的解析

在应用中,可以用两种形式来使用 Intent。

1) 显式 Intent

显式 Intent 指定了 Component 属性的 Intent(调用 setComponent(ComponentName)或者 setClass(Context,Class)来指定)。通过指定具体的组件类,通知应用启动对应的组件。

2) 隐式 Intent

隐式 Intent 没有指定 Comonent 属性的 Intent。这些 Intent 需要包含足够的信息,这样系统才能根据这些信息在所有可用的组件中,确定满足此 Intent 的组件。

对于直接 Intent，Android 不需要解析，因为目标组件已经很明确。Android 需要解析的是那些间接 Intent，通过解析将 Intent 映射给可以处理此 Intent 的 Activity、Service 或 Broadcast Receiver。

4. Intent 解析机制

Intent 解析机制主要通过查找已注册在 AndroidManifest.xml 中的所有＜intent-filter＞及其中定义的 Intent 来实现，通过 PackageManager（注：PackageManager 能够得到当前设备上所安装的 application package 的信息）查找能处理这个 Intent 的 Component。在这个解析过程中，Android 通过 Intent 的 Action、Type、Category 三个属性来进行判断，判断方法如下：

（1）如果 Intent 指明了 Action，则目标组件的 IntentFilter 的 Action 列表中就必须包含有这个 Action，否则不能匹配。

（2）如果 Intent 没有提供 Type，系统将从 Data 中得到数据类型。和 Action 一样，目标组件的数据类型列表中必须包含 Intent 的数据类型，否则不能匹配。

（3）如果 Intent 中的数据不是 content：类型的 URI，而且 Intent 也没有明确指定 Type，将根据 Intent 中数据的 scheme（比如 http：或者 mailto：）进行匹配。因此，Intent 的 scheme 必须出现在目标组件的 scheme 列表中。

（4）如果 Intent 指定了一个或多个 Category，则这些类别必须全部出现在组件的类别列表中。比如 Intent 中包含了两个类别：LAUNCHER_CATEGORY 和 ALTERNATIVE_CATEGORY，解析得到的目标组件必须至少包含这两个类别。

10.4 添加商户信息流程控制

下面介绍添加商户信息的具体实现方法。

1. 添加商户信息界面

添加商户信息界面如图 10.3 所示。

2. 添加商户信息功能流程的控制

1）AddShopFragment 类图

界面完成后，需要实现各组件数据的提取以及数据的验证，将商户信息传递给服务端，这个过程是在布局文件对应的 Activity 或者 Fragment 中完成的。

商户信息成功添加后，跳转页面需要根据具体业务需求以及用户体验综合考虑，确定后面的步骤。

比如，现在最好让商户对比之前填写的信息是否正确，同时预览商户页面，因此可跳转到商户详情界面。

如果需要用户整体的详细信息，就可以实现商户

图 10.3 添加商户信息界面

信息中心单独的页面,否则可直接转到所有商户列表等。

AddShopFragment.java 类图如图 10.4 所示。

AddShopFragment		
- rootView	: View	
- popupWindow	: PopupWindow	= null
- popupWindow_view	: View	= null
- shopType	: String	= "1"
- shopName	: String	
- shopDesc	: String	= ""
- shopTime	: String	
- shopAddress	: String	
- shopSpend	: String	
- shopLatitude	: String	
- shopLongitude	: String	
- shopPhone	: String	
- shopImage	: String	= null
- picFile	: File	
- photoUri	: Uri	
- filePath	: String	= null
- activity_result_camara_with_data	: int	= 1006
- activity_result_cropimage_with_data	: int	= 1007
+ newInstance ()		: AddShopFragment
+ <<Override>> onCreate (Bundle savedInstanceState)		: void
+ <<Override>> onCreateView (LayoutInflater inflater, ViewGroup container, Bundle savedInstanceState)		: View
+ onActivityCreated (Bundle savedInstanceState)		: void
+ onResume ()		: void
+ <<Override>> sendMessage (int opt)		: void
+ <<Override>> getData (int opt, Message msg)		: Object
+ onClick (View v)		: void
- getPopupWindow (int layoutId)		: void
- initPopWindow (int layoutId)		: void
# doTakePhoto ()		: void
# doCropPhoto ()		: void
- getCropImageIntent ()		: Intent
- cropImageUriByTakePhoto ()		: void
+ result (int requestCode, int resultCode, Intent data)		: void

图 10.4 AddShopFragment.java 类图

2) 判断商户信息是否符合规则

录入信息具有一定的规范,比如店铺名称的字数限制在什么范围,价格是否有小数点,电话号码是否匹配等。要实现这些功能,正则表达式就显得尤为重要了。正则表达式在 10.4.1 节有详细介绍。

对于这些限制,在 EditText 中,可以设置输入内容类型的属性 InputType,在此可以初步设置输入类型。若需详细设置,则要通过正则表达式在 Activity 中实现。

比如,当 EditText 需要输入密码时,需要将 InputType 设置为 android:inputType = "textPassword"。

在录入商铺的信息时,可对此进行简单判断,只需判断输入是否在指定长度内即可。实际运用中,会涉及具体的业务细节要求,因此需要更加详细地去验证。

代码如下:

```
//检查店铺名
EditText shopEdit = (EditText) rootView.findViewById(R.id.edit_add_shop_name);
shopName = shopEdit.getText() + "";
```

```java
if (shopName.length()<2)
{
    WidgetTools.setTVError(shopEdit, this.getResources().getString(R.string.toast_shop_name_tips), getActivity());
    return;
}
//检查店铺名
EditText descEdit = (EditText) rootView.findViewById(R.id.edit_add_shop_desc);
shopDesc =descEdit.getText() +"";
if (shopDesc.length()<10)
{
    WidgetTools.setTVError(descEdit, this.getResources().getString(R.string.toast_shop_desc_tips), getActivity());
    return;
}
//营业时间
EditText timeEdit = (EditText) rootView.findViewById(R.id.add_shop_time);
shopTime =timeEdit.getText() +"";
if (shopTime.length()<4)
{
    WidgetTools.setTVError(timeEdit, this.getResources().getString(R.string.toast_shop_time_tips), getActivity());
    return;
}
//店铺地址
EditText addrEdit = (EditText) rootView.findViewById(R.id.add_shop_addr);
shopAddress =addrEdit.getText() +"";
if (shopAddress.length()<4)
{
    WidgetTools.setTVError(addrEdit, this.getResources().getString(R.string.toast_shop_addr_tips), getActivity());
    return;
}
//平均花费
EditText speedEdit = (EditText) rootView.findViewById(R.id.add_shop_speed);
shopSpend =speedEdit.getText() +"";
if (shopSpend.length()<2)
{
    WidgetTools.setTVError(speedEdit, this.getResources().getString(R.string.toast_shop_speed_tips), getActivity());
    return;
}
//经度
```

```java
EditText latEdit =(EditText) rootView.findViewById(R.id.add_shop_lat);
shopLatitude =latEdit.getText()+"";
if (!NumberUtil.isDouble(shopLatitude))
{
    WidgetTools.setTVError(latEdit, this.getResources().getString(R.string.toast_shop_map_tips), getActivity());
    return;
}
//纬度
EditText longEdit =(EditText) rootView.findViewById(R.id.add_shop_long);
shopLongitude =longEdit.getText()+"";
if (!NumberUtil.isDouble(shopLongitude))
{
    WidgetTools. setTVError (longEdit, this.getResources().getString(R.string.toast_shop_map_tips), getActivity());
    return;
}
//电话
EditText phoneEdit =(EditText) rootView.findViewById(R.id.add_shop_phone);
shopPhone =phoneEdit.getText()+"";
if (shopPhone.length()<10)
{
    WidgetTools. setTVError (phoneEdit, this.getResources().getString(R.string.toast_shop_phone_tips), getActivity());
    return;
}
if (shopImage ==null)
{
    WidgetTools.setTVError(((Button) rootView.findViewById(R.id.button_choice_pic)), this.getResources().getString(R.string.toast_shop_img_url_tips), getActivity());
    return;
}
sendMessage(OPT.POST_SHOP);
```

数据全部验证通过后，发送信息到服务端。代码如下：

```java
param.put("act", "postShop");
param.put("shopAddress", shopAddress);
param.put("shopDesc", shopDesc);
param.put("shopImage", shopImage);
param.put("shopLatitude", shopLatitude);
param.put("shopLongitude", shopLongitude);
param.put("shopName", shopName);
param.put("shopPhone", shopPhone);
```

```
param.put("shopSpend", shopSpend);
param.put("shopTime", shopTime);
param.put("shopType", shopType);
```

3）选择并裁剪图片

上传商户图片时，可以从相册或者 SD 卡内选择图片上传，也可以使用相机立即拍照上传。为了规范图片的大小与质量，需要对图片宽、高进行设定，由商户自己进行裁剪。

在 SD 卡内的 meDemo 文件目录中，创建子目录 upload，将裁剪后的图片临时存放于此处。代码如下：

```
File uploadFileDir = new File (Environment.getExternalStorageDirectory(),
"meDemo" + File.separator + "upload");
if (!uploadFileDir.exists())
{
    uploadFileDir.mkdirs();
}
```

调用图片裁剪程序，进入图片管理界面，代码如下：

```
final Intent intent = getCropImageIntent();
```

从图片选择到图片裁剪的方法，代码如下：

```
protected void doCropPhoto()
{
    try
    {
        File uploadFileDir = new File(Environment.getExternalStorageDirectory(),
"meDemo" + File.separator + "upload");
        if (!uploadFileDir.exists())
        {
            uploadFileDir.mkdirs();
        }
        //Create a media file name
        String timeStamp = new SimpleDateFormat("yyyyMMdd_HHmmss", Locale.
getDefault()).format(new Date());
        picFile = new File(uploadFileDir, "img" + timeStamp + ".jpg");
        if (!picFile.exists())
        {
            picFile.createNewFile();
        }
        photoUri = Uri.fromFile(picFile);
        final Intent intent = getCropImageIntent();
        ((MainActivity) getActivity()).startActivityForResult(intent, activity_
result_cropimage_with_data);
    } catch (Exception e)
```

```
        {
            e.printStackTrace();
        }
    }
```

Intent.ACTION_PICK 根据传入的文件类型，设置如下参数：

setType：设置文件类型。

"crop"：是否裁剪。

"aspectX"/"aspectY"：裁剪框比例。

"outputX"/"outputY"：设置图片的宽、高像素。

"noFaceDetection"：脸部识别。

"scale"：是否支持缩放。

"return-data"：是否直接返回二进制数据。

MediaStore.EXTRA_OUTPUT：文件设置。

"outputFormat"：输出流编码，可以设置图片编码的 JPEG、PNG 等格式。

代码如下：

```java
private Intent getCropImageIntent()
{
    Intent intent = new Intent(Intent.ACTION_PICK, photoUri);
    intent.setType("image/*");
    intent.putExtra("crop", "true");
    intent.putExtra("aspectX", 1);
    intent.putExtra("aspectY", 1);
    intent.putExtra("outputX", 320);
    intent.putExtra("outputY", 320);
    intent.putExtra("noFaceDetection", true);
    intent.putExtra("scale", true);
    intent.putExtra("return-data", false);
    intent.putExtra(MediaStore.EXTRA_OUTPUT, photoUri);
    intent.putExtra("outputFormat", Bitmap.CompressFormat.JPEG.toString());
    intent.setFlags(Intent.FLAG_ACTIVITY_SINGLE_TOP);
    return intent;
}
```

4）照相机拍照

调用照相机拍照与直接选择图片裁剪有些许差异。首先要调用相机拍照，传入 Intent 参数 MediaStore.ACTION_IMAGE_CAPTURE 来启动相机。代码如下：

```java
protected void doTakePhoto()
{
    try
    {
        File uploadFileDir = new File(Environment.getExternalStorageDirectory(),
```

```
            "meDemo" +File.separator +"upload" +File.separator);
        Intent cameraIntent =new Intent(MediaStore.ACTION_IMAGE_CAPTURE);
        if (!uploadFileDir.exists())
        {
            uploadFileDir.mkdirs();
        }
        //Create a media file name
        String timeStamp =new SimpleDateFormat("yyyyMMdd_HHmmss", Locale.
getDefault()).format(new Date());
        picFile =new File(uploadFileDir, "img" +timeStamp +".jpg");
        if (!picFile.exists())
        {
            picFile.createNewFile();
        }
        photoUri =Uri.fromFile(picFile);
        cameraIntent.putExtra(MediaStore.EXTRA_OUTPUT, photoUri);
        cameraIntent.setFlags(Intent.FLAG_ACTIVITY_SINGLE_TOP);
        ((MainActivity) getActivity()).startActivityForResult(cameraIntent,
activity_result_camara_with_data);
    } catch (ActivityNotFoundException e)
    {
        e.printStackTrace();
    } catch (IOException e)
    {
        e.printStackTrace();
    }
}
```

拍完照后,调用 Activity 的 result 方法启动程序裁剪程序 cropImageUriByTakePhoto(),参数与选择图片进行裁剪 Intent 传入参数有些差别。请注意"com. android. camera. action. CROP"和 setDataAndType(photoUri,"image/ * ")的不同之处。

代码如下:

```
private void cropImageUriByTakePhoto()
{
    Intent intent =new Intent("com.android.camera.action.CROP");
    intent.setDataAndType(photoUri, "image/*");
    intent.putExtra("crop", "true");
    intent.putExtra("aspectX", 1);
    intent.putExtra("aspectY", 1);
    intent.putExtra("outputX", 320);
    intent.putExtra("outputY", 320);
    intent.putExtra("scale", true);
    intent.putExtra(MediaStore.EXTRA_OUTPUT, photoUri);
```

```
        intent.putExtra("return-data", false);
        intent.putExtra("outputFormat", Bitmap.CompressFormat.JPEG.toString());
        intent.putExtra("noFaceDetection", true); //no face detection
        intent.setFlags(Intent.FLAG_ACTIVITY_SINGLE_TOP);
        ((MainActivity) getActivity()).startActivityForResult(intent, activity_
result_cropimage_with_data);
    }
```

5）接收系统截图返回数据

调用系统的截图工具。截图完成后，onActivityResult 会返回截图后的相关数据。根据返回的 code 可判断是哪个步骤处理完成后返回的数据。代码如下：

```
    public void result(int requestCode, int resultCode, Intent data)
    {
        switch (requestCode)
        {
            case activity_result_camara_with_data://拍照
                try
                {
                    cropImageUriByTakePhoto();
                } catch (Exception e)
                {
                    e.printStackTrace();
                }
                break;
            case activity_result_cropimage_with_data:
                try
                {
                    if (photoUri !=null)
                    {
                        sendMessage(OPT.UPLODA_IMG);
                    }
                } catch (Exception e)
                {
                    e.printStackTrace();
                }
                break;
        }
    }
```

onActivityReuslt 是 Activity 的函数。在 Fragment 中没有这个回调函数，因此只能通过装载 Fragment 的父类 Activity MainActivity 来中转这个方法。代码如下：

```
        ((MainActivity) getActivity()).startActivityForResult(cameraIntent,
activity_result_camara_with_data);
```

父类 MainActivity 得到数据后,上传 Fragment 自定义的 result 方法,间接实现数据传递,完成截图功能。代码如下:

```
@Override
protected void onActivityResult(int requestCode, int resultCode, Intent data)
{
    if (resultCode !=RESULT_OK || resultCode ==RESULT_CANCELED)
        return;
    switch (requestCode)
    {
        case activity_result_camara_with_data://拍照
        case activity_result_cropimage_with_data:
            addShopFragment.result(requestCode, resultCode, data);
            break;
    }
    super.onActivityResult(requestCode, resultCode, data);
}
```

6)上传文件到服务器

得到裁剪后的图片地址后,上传到服务端,首先需要封装数据,上传图片参数。代码如下:

```
filePath =BitmapUtil.savePic(picFile.getPath());
if (filePath ==null)
{
    if (null !=progressDialog && progressDialog.isShowing())
    {
        progressDialog.dismiss();
    }
    return;
}
param.put("act", "postImage");
param.put("file", filePath);
```

当然,这样简单处理肯定是无法传输文件的,还需要用到 appache 开源组织的上传文件 httpmime-4.2.1.jar 包。

com.me.demo.util.Tools 类是集中处理与服务端数据封装的工具类,可用方法 json2MultipartEntity(JSONObject json) 封装数据,当遇到 file 开头的参数时,new FileBody(avatar)封装这个路径的文件到组件中。代码如下:

```
public static MultipartEntity json2MultipartEntity(JSONObject json)
{
    MultipartEntity mulentity =new MultipartEntity(HttpMultipartMode.BROWSER
_COMPATIBLE);
    Iterator<String>iterator =json.keys();
```

```java
    while (iterator.hasNext())
    {
        try
        {
            String key = iterator.next();
            String value = json.getString(key);
            try
            {
                if (key.startsWith("file"))
                {
                    File avatar = new File(value);
                    mulentity.addPart("file[]", new FileBody(avatar));
                } else
                {
                    mulentity.addPart(key, new StringBody(value, Charset.forName(HTTP.UTF_8)));
                }
            } catch (UnsupportedEncodingException e)
            {
                e.printStackTrace();
            }
        } catch (JSONException e)
        {
            e.printStackTrace();
        }
    }
    return mulentity;
}
```

封装完毕,可用方法 uploadImg(HttpPost post)传递到服务端。代码如下:

```java
public static String uploadImg(HttpPost post)
{
    HttpClient httpclient = new DefaultHttpClient();
    String resultStr = null;
    HttpResponse response;
    try
    {
        response = httpclient.execute(post);
        HttpEntity entity = response.getEntity();
        byte[] result = EntityUtils.toByteArray(entity);
        if (response != null)
            resultStr = new String(result, "utf-8");
    } catch (ClientProtocolException e)
    {
```

```
            e.printStackTrace();
        } catch (IOException e)
        {
            e.printStackTrace();
        }
        return resultStr;
    }
```

10.5 知识点回顾与技能扩展

10.5.1 知识点回顾

本章主要知识点如下：
(1) 系统自带截图的理解和使用。
(2) Intent 和 Intent Action 的理解与常用标志的使用。
(3) 正则表达式的理解与使用。

10.5.2 技能扩展

1. 正则表达式简介

正则表达式是对字符串操作的一种逻辑公式，就是用事先定义好的一些特定字符及这些特定字符的组合，组成一个"规则字符串"。这个"规则字符串"用来表达对字符串的一种过滤逻辑。

给定一个正则表达式和另一个字符串，我们可以达到如下目的：
(1) 判断给定的字符串是否符合正则表达式的过滤逻辑（称作"匹配"）。
(2) 通过正则表达式，可以从字符串中获取我们想要的特定部分。

正则表达式的特点是：
(1) 灵活性、逻辑性和功能性非常强。
(2) 可以用极简单的方式实现对字符串的复杂控制。
(3) 对于刚接触的人来说，比较晦涩难懂。

由于正则表达式的主要应用对象是文本，因此它可应用在各种文本编辑器中，小到著名编辑器 EditPlus，大到 Microsoft Word、Visual Studio 等大型编辑器，都可以使用正则表达式来处理文本内容。

下面介绍正则表达式的语法。

(1) 句点符号：代表任意符号。例如，搜索一个包含字符"me"的字符串，搜索用的正则表达式就是"me"。搜索单词以字母"m"开头，以字母"e"结束，这样完整的正则表达式就是"m. e"。

(2) 方括号：提供选择的符号。例如，正则表达式 m[i]e，等同于 mie。若变换为

m[if]e,则中间位的字符只能为 i 或者 f 才匹配。

（3）"或"符号"|"与圆括号"()"。例如，正则表达式 m(o|1|23)e,其匹配式只能是 moe、m1e、m23e 中的一个。显然，"|"符号进行的是"或"运算，而且必须与"()"配合使用。

（4）匹配次数的符号如表 10.1 所示，这些符号用来确定紧靠该符号左边的符号出现的次数。

表 10.1 匹配次数的符号

符 号	次 数	符 号	次 数
*	0 次或多次	{n}	恰好 n 次
+	1 次或多次	{n,m}	从 n 次到 m 次
?	0 次或 1 次		

（5）"否"符号。"^"符号称为"否"符号。如果用在方括号内，则"^"表示不想匹配的字符。

（6）圆括号与空白符号。"\s"符号是空白符号，匹配所有的空白字符，包括 Tab 字符。

（7）表 10.2 所示为常见正则表达式创建的快捷符号。

表 10.2 快捷符号

符 号	等价的正则表达式	符 号	等价的正则表达式
\d	[0-9]	\W	[^A-Z0-9]
\D	[^0-9]	\s	[\t\n\r\f]
\w	[A-Z0-9]	\S	[^\t\n\r\f]

2. 正则表达式实例

1）匹配手机号

```
Pattern p =Pattern.compile("^((13[0-9])|(15[^4,\\D])|(18[0-9]))\\d{8}$");
Matcher m =p.matcher("15881049999");
m.matches();
```

说明：^表示行开始，[0-9]表示在 0～9 中取一个数，15[^4,\\D]表示在 0～9 中除 4 以外的数字，\\{8}表示后面还有一个数字。

2）匹配用户名

```
Pattern p =Pattern.compile("^[a-z0-9_-]{3,15}$");
Matcher m =p.matcher("fuiden321");
m.matches();
```

说明：[a-z0-9_-]表示匹配列表中的 a～z、0～9、下画线、连字符，{3,15}表示长度至少为 3 个字符，最大长度为 15。

3) 匹配密码

```
Pattern p =Pattern.compile("(?=.*\\d)(?=.*[a-z])(?=.*[A-Z])(?=.*[@#$%]).{6,20}");
Matcher m =p.matcher("fuiden321");
m.matches();
```

说明:

(# 组开始。

(?=.*\d) # 必须包含 0~9 中的一个数字。

(?=.*[a-z]) # 必须包含一个小写字母。

(?=.*[A-Z]) # 必须包含一个大写字母。

(?=.*[@#$%]) # 必须包含一个列表中的特殊字符"@#$%"。

. # 检查所有字符串与前面条件的匹配情况。

{6,20} # 长度至少为 6 个字符,最大长度为 20。

) # 组结束。

4) 匹配 E-mail

```
Pattern p =Pattern.compile("^[_A-Za-z0-9-]+(\\.[_A-Za-z0-9-]+)*@[A-Za-z0-9]+(\\.[A-Za-z0-9]+)*(\\.[A-Za-z]{2,})$");
Matcher m =p.matcher("fuiden321@qq.com");
m.matches();
```

说明:[_A-Za-z0-9-]+ # 必须以中括号中的字符为起始字符[],必须包含一个或多个(+)。

(# 组#1 开始。

\\.[_A-Za-z0-9-]+ # 接下来是一个点".",和中括号内的字符[],必须包含一个或者多个(+)。

)* # 组#1 结束,这个组是可选的(*)。

@ # 必须包含一个"@"符号。

[A-Za-z0-9]+ # 接下来是中括号内的字符[],必须包含一个或者多个(+)。

(# 组 #2 开始——一级 TLD 检查。

\\.[A-Za-z0-9]+ # 接下来是一个点".",和中括号内的字符[],必须包含一个或者多个(+)。

)* # 组#2 结束,这个组是可选的(*)。

(# 组#3 开始——二级 TLD 检查。

\\.[A-Za-z]{2,} # 接下来是一个点".",和中括号内的字符[],最小长度为 2。

) # 组#3 结束。

3. 常用 Intent 和 Intent Action

1) 常用的 Intent

Intent 和 Intent Action 关联着许多系统功能和系统标志,对于开发者而言,调用系统的功能可以大大降低开发难度，减少重复代码带来的冗余。因此掌握 Intent 的知识，尤为重要。

下面主要讲解一些最常用的功能，要了解更多知识，请参照本书附录 B 以及官方网站的文档。

（1）从 google 搜索内容。

设置 Intent.ACTION_WEB_SEARCH，调用 Google 搜索。

```
Intent intent = new Intent();
intent.setAction(Intent.ACTION_WEB_SEARCH);
intent.putExtra(SearchManager.QUERY, "searchString");
startActivity(intent);
```

（2）浏览网页。

传入 Intent.ACTION_VIEW，uri 设置一个网页地址。启动 Intent，系统会自动判断调用浏览器，打开网页地址。

```
Uri uri = Uri.parse("http://www.google.com");
Intent it = new Intent(Intent.ACTION_VIEW, uri);
startActivity(it);
```

（3）显示地图。

传入 Intent.ACTION_VIEW，uri 设置为经纬度坐标。启动 Intent，系统会自动判断启动地图应用。

```
Uri uri = Uri.parse("geo:38.899533,-77.036476");
Intent it = new Intent(Intent.ACTION_VIEW, uri);
startActivity(it);
```

（4）路径规划。

传入 Intent.ACTION_VIEW，uri 提供路径规划参数。启动 Intent，系统会自动判断启动地图路径规划功能。

```
Uri uri = Uri.parse("http://maps.google.com/maps?f=dsaddr=startLat%20startLng&daddr=endLat%20endLng&hl=en");
Intent it = new Intent(Intent.ACTION_VIEW, uri);
startActivity(it);
```

（5）拨打电话。

传入 Intent.ACTION_DIAL，uri 传入电话号码。启动 Intent，系统打开拨号界面，并录入电话号码。

```
Uri uri = Uri.parse("tel:xxxxxx");
Intent it = new Intent(Intent.ACTION_DIAL, uri);
startActivity(it);
```

(6) 调用发短信的程序。

方式一：传入 Intent.ACTION_VIEW，关键是设置 Type 类型为"vnd.android-dir/mms-sms"，并在"sms_body"后面录入短信的内容。

```
Intent it =new Intent(Intent.ACTION_VIEW);
it.putExtra("sms_body", "The SMS text");
it.setType("vnd.android-dir/mms-sms");
startActivity(it);
```

方式二：传入 Intent.ACTION_SENDTO，只需要在"sms_body"后面录入短信的内容。

```
Uri uri =Uri.parse("smsto:0800000123");
Intent it =new Intent(Intent.ACTION_SENDTO, uri);
it.putExtra("sms_body", "The SMS text");
startActivity(it);
```

(7) 发送彩信。

彩信与短信的区别在于，彩信可以传送图片文件，所以需要写入流。在传入 Intent.ACTION_SENDTO 的基础上，传入 uri。传入图片地址并设置 Type 类型为"image/png"。

```
Uri uri =Uri.parse("content://media/external/images/media/23");
Intent it =new Intent(Intent.ACTION_SEND);
it.putExtra("sms_body", "some text");
it.putExtra(Intent.EXTRA_STREAM, uri);
it.setType("image/png");
startActivity(it);
```

(8) 发送 E-mail。

方式一：传入 Intent.ACTION_SENDTO，Uri 传入 E-mail 地址。

```
Uri uri =Uri.parse("mailto:xxx@abc.com");
Intent it =new Intent(Intent.ACTION_SENDTO, uri);
startActivity(it);
```

方式二：传入 Intent.ACTION_SENDTO，设置 Type 为"text/plain"，调用 Intent.createChooser 方法。E-mail 收件地址作为 Extra 传入。

```
Intent it =new Intent(Intent.ACTION_SEND);
it.putExtra(Intent.EXTRA_EMAIL, "me@abc.com");
it.putExtra(Intent.EXTRA_TEXT, "The email body text");
it.setType("text/plain");
startActivity(Intent.createChooser(it, "Choose Email Client"));
```

方式三：没有附件的文本 E-mail。传入 Intent.ACTION_SENDTO，设置 Type 为"message/rfc822"，调用 Intent.createChooser 方法。

```
Intent it=new Intent(Intent.ACTION_SEND);
String[] tos={"me@abc.com"};
String[] ccs={"you@abc.com"};
it.putExtra(Intent.EXTRA_EMAIL, tos);
it.putExtra(Intent.EXTRA_CC, ccs);
it.putExtra(Intent.EXTRA_TEXT, "The email body text");
it.putExtra(Intent.EXTRA_SUBJECT, "The email subject text");
it.setType("message/rfc822");
startActivity(Intent.createChooser(it, "Choose Email Client"));
```

方式四：含有附件的 E-mail。传入 Intent.ACTION_SENDTO，设置 Type 为 "audio/mp3"等，调用 Intent.createChooser 方法。

```
Intent it =new Intent(Intent.ACTION_SEND);
it.putExtra(Intent.EXTRA_SUBJECT, "The email subject text");
it.putExtra(Intent.EXTRA_STREAM, "file:///sdcard/mysong.mp3");
it.setType("audio/mp3");
startActivity(Intent.createChooser(it, "Choose Email Client"));
```

(9) 播放多媒体。

方式一：传入 Intent.ACTION_VIEW，传入有音频或视频文件名的 Uri，setDataAndType 为 audio/mp3。

```
Intent it =new Intent(Intent.ACTION_VIEW);
Uri uri =Uri.parse("file:///sdcard/song.mp3");
it.setDataAndType(uri, "audio/mp3");
startActivity(it);
```

方式二：传入 Intent.ACTION_VIEW，传入 Uri，设置为 Media.INTERNAL_CONTENT_URI。

```
Uri uri =Uri.withAppendedPath(MediaStore.Audio.Media.INTERNAL_CONTENT_URI, "地址");
Intent it =new Intent(Intent.ACTION_VIEW, uri);
startActivity(it);
```

(10) 卸载或安装 apk。

卸载 apk：传入 Intent.ACTION_DELETE，Uri 传入"package"的值为目标 apk 的包名。

```
Uri uri =Uri.fromParts("package", strPackageName, null);
Intent it =new Intent(Intent.ACTION_DELETE, uri);
startActivity(it);
```

安装 apk：传入 Intent.ACTION_PACKAGE_ADDED，Uri 传入"package"的值为目标 apk 的地址。

```
Uri installUri =Uri.fromParts("package", "xxx", null);
Intent returnIt =new Intent(Intent.ACTION_PACKAGE_ADDED, installUri);
startActivity(returnIt);
```

(11) 打开照相机。

方式一：发送广播的形式，启动系统的照相机。

```
Intent i =new Intent(Intent.ACTION_CAMERA_BUTTON, null);
getActivity().sendBroadcast(i);
```

方式二：直接打开某个路径的图片。

```
String fileName ="test";
ContentValues values =new ContentValues();
values.put(Images.Media.TITLE, fileName);
values.put("_data", fileName);
values.put(Images.Media.PICASA_ID, fileName);
values.put(Images.Media.DISPLAY_NAME, fileName);
values.put(Images.Media.DESCRIPTION, fileName);
values.put(Images.ImageColumns.BUCKET_DISPLAY_NAME, fileName);
Uri photoUri =getActivity().getContentResolver().insert(MediaStore.Images.
Media.EXTERNAL_CONTENT_URI, values);
```

方式三：选择图片，返回图片地址，并由 Activity 提供回调函数 onActivityResult()。

```
Intent inttPhoto =new Intent(MediaStore.ACTION_IMAGE_CAPTURE);
inttPhoto.putExtra(MediaStore.EXTRA_OUTPUT, photoUri);
startActivityForResult(inttPhoto, 10);
```

(12) 从 gallery 选取图片。

传入 Intent.ACTION_GET_CONTENT。

```
Intent i =new Intent();
i.setType("image/*");
i.setAction(Intent.ACTION_GET_CONTENT);
startActivityForResult(i, 11);
```

(13) 打开录音机。

传入 Media.RECORD_SOUND_ACTION。

```
Intent mi =new Intent(Media.RECORD_SOUND_ACTION);
startActivity(mi);
```

(14) 显示应用详细列表。

传入 Intent.ACTION_VIEW，Uri 传入接口查看应用商城某个应用的详情页。

```
Uri uri =Uri.parse("market://details?id=<packagename>");
Intent it =new Intent(Intent.ACTION_VIEW, uri);
startActivity(it);
```

(15) 寻找应用。

传入 Intent.ACTION_VIEW，Uri 传入接口调用系统的应用市场功能。

```
Uri uri =Uri.parse("market://search?q=pname:pkg_name");
Intent it =new Intent(Intent.ACTION_VIEW, uri);
startActivity(it);
```

(16) 打开联系人列表。

方式一：通过 MIME 类型打开联系人管理工具。

```
final int INTENT_TAG =123;
Intent i =new Intent();
i.setAction(Intent.ACTION_GET_CONTENT);
i.setType("vnd.android.cursor.item/phone");
startActivityForResult(i, INTENT_TAG);
```

方式二：通过 Uri 类型打开联系人管理工具。

```
final int INTENT_TAG =123;
Uri uri =Uri.parse("content://contacts/people");
Intent it =new Intent(Intent.ACTION_PICK, uri);
startActivityForResult(it, INTENT_TAG);
```

(17) 打开其他程序。

打开其他程序，导入对应的 action＝"android.intent.action.MAIN"，就可以实现启动其他程序的功能。

```
final int RESULT_OK =1234;
Intent i =new Intent();
ComponentName cn = new ComponentName("com.yellowbook.android2","com.yellowbook.android2.AndroidSearch");
i.setComponent(cn);
i.setAction("android.intent.action.MAIN");
startActivityForResult(i, RESULT_OK);
```

2) 常用的 Intent Action

android.intent.action.ALL_APPS
android.intent.action.ANSWER
android.intent.action.ATTACH_DATA
android.intent.action.BUG_REPORT
android.intent.action.CALL
android.intent.action.CALL_BUTTON
android.intent.action.CHOOSER
android.intent.action.CREATE_LIVE_FOLDER
android.intent.action.CREATE_SHORTCUT

android.intent.action.DELETE
android.intent.action.DIAL
android.intent.action.EDIT
android.intent.action.GET_CONTENT
android.intent.action.INSERT
android.intent.action.INSERT_OR_EDIT
android.intent.action.MAIN
android.intent.action.MEDIA_SEARCH
android.intent.action.PICK
android.intent.action.PICK_ACTIVITY
android.intent.action.RINGTONE_PICKER
android.intent.action.RUN
android.intent.action.SEARCH
android.intent.action.SEARCH_LONG_PRESS
android.intent.action.SEND
android.intent.action.SENDTO
android.intent.action.SET_WALLPAPER
android.intent.action.SYNC
android.intent.action.SYSTEM_TUTORIAL
android.intent.action.VIEW
android.intent.action.VOICE_COMMAND
android.intent.action.WEB_SEARCH
android.net.wifi.PICK_WIFI_NETWORK
android.settings.AIRPLANE_MODE_SETTINGS
android.settings.APN_SETTINGS
android.settings.APPLICATION_DEVELOPMENT_SETTINGS
android.settings.APPLICATION_SETTINGS
android.settings.BLUETOOTH_SETTINGS
android.settings.DATA_ROAMING_SETTINGS
android.settings.DATE_SETTINGS
android.settings.DISPLAY_SETTINGS
android.settings.INPUT_METHOD_SETTINGS
android.settings.INTERNAL_STORAGE_SETTINGS
android.settings.LOCALE_SETTINGS
android.settings.LOCATION_SOURCE_SETTINGS
android.settings.MANAGE_APPLICATIONS_SETTINGS
android.settings.MEMORY_CARD_SETTINGS
android.settings.NETWORK_OPERATOR_SETTINGS

android. settings. QUICK_LAUNCH_SETTINGS
android. settings. SECURITY_SETTINGS
android. settings. SETTINGS
android. settings. SOUND_SETTINGS
android. settings. SYNC_SETTINGS
android. settings. USER_DICTIONARY_SETTINGS
android. settings. WIFI_IP_SETTINGS
android. settings. WIFI_SETTINGS
android. settings. WIRELESS_SETTINGS

第 11 章 让用户使用体验更佳

11.1 用户手机网络环境

用户手机网络存在多种环境，GSM、CDMA、3G、4G 都是运营商提供的有偿使用的网络环境，还有一种不限流量的 Wi-Fi 环境。

如果要看电影，那么使用哪种环境呢？首选 Wi-Fi 模式看电影，心情可能更舒畅。

如果是在非 Wi-Fi 环境，不下载缓冲，就需要得到当前链接的网络模式。

```
/**
 * @param context
 * @return 0:wifi 1:mobile 2:默认 -1:没有发现网络连接
 */
public static int getNetType(Context context)
{
    //获取连接管理器
    ConnectivityManager mConnect = (ConnectivityManager) context.getSystemService(Context.CONNECTIVITY_SERVICE);
    //创建一个网络信息对象
    NetworkInfo mNetInfo =mConnect.getActiveNetworkInfo();
    if (mNetInfo !=null && mNetInfo.isAvailable())
    {
        int netType =mNetInfo.getType();
        if (netType ==ConnectivityManager.TYPE_MOBILE)
        {
            return 0;
        } else if (netType ==ConnectivityManager.TYPE_WIFI)
        {
            return 1;
        } else
        {
            return 2;
        }
    } else
```

```
    {
        return -1;
    }
}
```

获取网络类型后,再根据具体需求做出相应的处理。比如在非 Wi-Fi 环境下,不显示图片;在 Wi-Fi 环境下,要显示图片以及控制的开关等。

11.2 知识点回顾

本章主要知识点如下:
(1) ConnectivityManager 的理解与使用。
(2) 根据对业务的理解来设定 Wi-Fi 与操作的关系。

第 12 章 发布和管理 Android 应用

12.1 为何要发布

一个应用程序完成后,就需要面对用户了。用户要安装 APK,首先就需要下载这个 APK 应用程序。因此第一个问题就是用户在哪里下载这个 APK 应用程序。通常,首先想到的是官网是否提供下载,因此需要在官网提供一个下载通道给用户。有的用户习惯使用某一个应用市场,下载 APK 应用时首先想到的就是到这个应用市场去搜索。要尽量满足客户的需求,使应用尽可能覆盖用户的使用习惯,就要将应用发布到第三方市场。

12.2 在哪里发布

APK 通常有两种发布方式,分别为发布到官网及第三方市场。

发布 APK 到官网的步骤如下:
(1) 将 APK 文件放到服务端。
(2) 将 APK 下载链接与官网的下载链接绑定。

12.3 如何发布到第三方市场

12.3.1 在 Eclipse 中对 Android 应用签名

(1) 在 Eclipse 工程中右击工程,在弹出的选项中选择 Android Tools→Export Signed Application Package,如图 12.1 所示。
(2) 选择需要打包的 Android 项目工程,如图 12.2 所示。
(3) 如果已有私钥文件,则选择私钥文件并输入密码。如果没有私钥文件,则需参见第(6)、(7)步创建私钥文件,如图 12.3 所示。
(4) 输入私钥别名和密码,如图 12.4 所示。

第12章 发布和管理 Android 应用

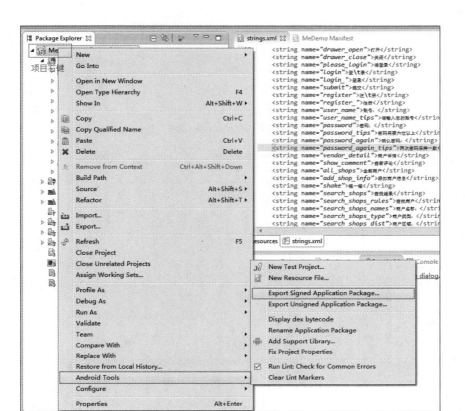

图 12.1　Eclipse 签名打包菜单

图 12.2　选择打包的项目工程

图 12.3 选择私钥并输入密码

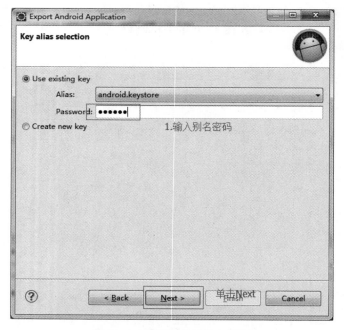

图 12.4 选择别名并输入密码

（5）选择 APK 存储的位置，单击 Finish 按钮完成设置，开始生成，如图 12.5 所示。

图 12.5　选择 APK 存放位置

（6）没有私钥文件时，创建私钥文件，有两种方式。

方式一：创建私钥，填入私钥信息，如图 12.6、图 12.7 所示。

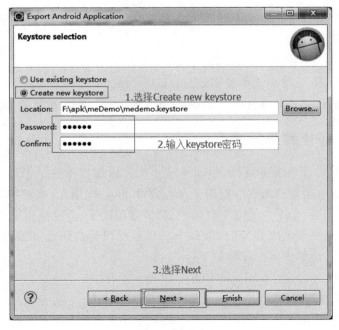

图 12.6　创建私钥

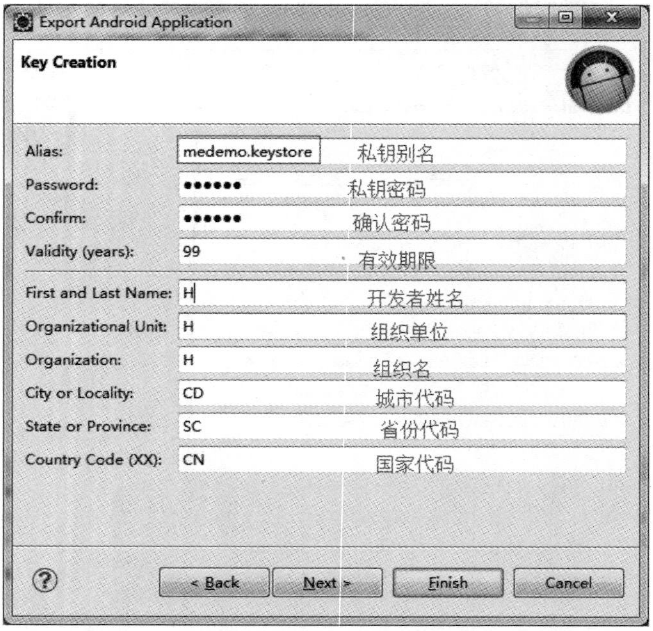

图 12.7 填入私钥信息

方式二：

在 DOS 中进入 JDK 的 bin 目录。

运行如下命令：

```
keytool -genkey -alias android.keystore -keyalg RSA -validity 20000 -keystore android.keystore
```

其中,-validity 20000 代表有效期天数。

命令完成后,bin 目录中会生成 android.keystore。

查看命令 keytool -list -keystore "android.keystore",输入设置的 keystore 密码。

12.3.2 发布 APK 到第三方市场

Android 项目由于开源的特性,市场不像苹果商店那样唯一。市面上的 Android 应用市场非常多,其中主流的第三方应用平台有安卓市场、应用汇、安智市场、N 多市场、豌豆荚、机锋市场、91 手机助手、360 手机助手、腾讯应用助手、百度应用助手等。

所有平台都需要申请成为开发者,每个平台需要填写的信息和发布流程是不同的。下面以豌豆荚为例简单讲解。步骤如下：

（1）打开主页 http://open.wandoujia.com/home。

（2）注册用户,如图 12.8 所示。

（3）注册并激活成功后,再进入主页 http://open.wandoujia.com/home,同意开发者协议,如图 12.9 所示。

图 12.8 注册用户

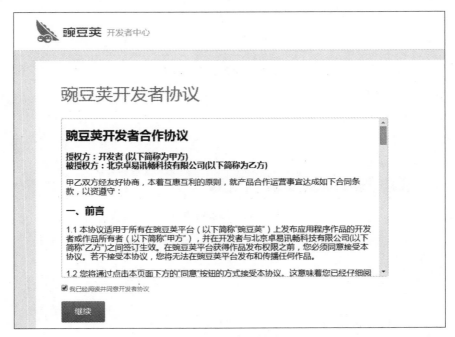

图 12.9 同意开发者协议

(4) 填写开发者信息,是个人还是企业,如图 12.10 所示。
(5) 审核通过后,选择"应用"→"添加新应用",如图 12.11 所示。
(6) 提交审核,上传安装包,如图 12.12 所示。
(7) 审核通过后,应用成功上架,在豌豆荚市场就发布成功了。

图 12.10 填写开发者信息

图 12.11 添加新应用

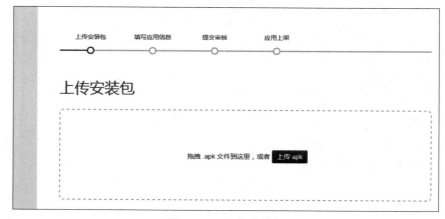

图 12.12 上传安装包

12.4 版本与版本管理

12.4.1 设置版本号和版本名

在 AndroidManifest.xml 中设置版本号和版本名。代码如下：

```
<manifest xmlns:android="http://schemas.android.com/apk/res/android"
    package="com.me.demo"
    android:versionCode="1"
    android:versionName="1.0" >
```

android：versionCode＝"1"设置的是版本号；android：versionName＝"1.0"设置的是版本名。

12.4.2 获取当前版本信息

获取当前版本信息的代码如下：

```
PackageManager packageManager =getPackageManager();
PackageInfo packInfo =null;
try
{
    String packageName =getPackageName();
    packInfo =packageManager.getPackageInfo(packageName, 0);
    int code =packInfo.versionCode;
    String name =packInfo.packageName;
} catch (PackageManager.NameNotFoundException e)
{
    e.printStackTrace();
}
```

packInfo.versionCode 得到 AndroidManifest.xml 设置的 versionCode；packInfo.packageName 得到 AndroidManifest.xml 设置的 android：versionName。

12.5 如何让用户升级

12.5.1 服务器准备

1. 数据库准备

客户端版本如表 12.1 所示。

表 12.1 客户端版本

名称	说明	数据类型	主键/外键/非空
id	id	int	P
version_code	版本号	varchar	
version_content	版本内容	varchar	
version_url	APP下载地址	varchar	
plat	1：android，2：ios	int	

2. 接口准备

(1) 调用方式：POST。应用升级参数如表 12.2 所示。

表 12.2 应用升级参数

请求参数	必选	类型及范围	说明
act	Y	String	update
version_code	Y	String	当前版本code
plat	Y	String	1

(2) 返回方式：JSON。升级请求接口参数如表 12.3 所示。

表 12.3 升级请求接口参数

返回值字段	字段类型	字段说明
flag	Int	1：成功，0：失败
msg	String	消息提示信息
id	JSONObject	id
version_code	String	版本code
version_content	String	版本内容
version_url	String	APP下载地址

12.5.2 客户端实现

1. 请求接口数据封装

根据接口文档，将参数封装到 JSON 数据中，并传递给服务端。代码如下：

```
PackageInfo packageInfo = application.getPackageInfo();
param.put("version_code", packageInfo.versionCode);
param.put("act", "update");
param.put("plat", 1);
```

接口返回数据后,与当前 APP 的 version_code 进行比较,若当前的更大,则不需要更新,否则就需要更新。

```java
JSONObject update = new JSONObject(version);
int code = Integer.parseInt(update.getString("version_code"));
if (code <= application.getPackageInfo().versionCode)
{
    return null;
}
```

2. 更新版本流程控制

1) Android 代号与版本

Android 代号与版本对照如图 12.13 所示。

Android代号	Android版本
ECLAIR_0_1	Android 2.0.1
ECLAIR_MR1	Android 2.1
FROYO	Android 2.2
GINGERBREAD	Android 2.3
GINGERBREAD_MR1	Android 2.3.3
HONEYCOMB	Android 3.0
HONEYCOMB_MR1	Android 3.1
HONEYCOMB_MR2	Android 3.2
ICE_CREAM_SANDWICH	Android 4.0

图 12.13　Android 代号与版本对照

Android 系统自带下载器,这是 Android 3.0 以后才支持的,所以首先要判断当前手机版本是否大于 Android 3.0 版本。

```java
if (Build.VERSION.SDK_INT >= Build.VERSION_CODES.HONEYCOMB_MR1)
```

若 Build.VERSION.SDK_INT 大于或等于 Build.VERSION_CODES.HONEYCOMB_MR1,则调用系统下载器。代码如下:

```java
prefs = PreferenceManager.getDefaultSharedPreferences(getBaseContext());
downloadManager = (DownloadManager) getSystemService(Activity.DOWNLOAD_SERVICE);
BaseFragmentActivity.DL_ID = name;
Uri uri = Uri.parse(url);
DownloadManager.Request request = new DownloadManager.Request(uri);
request.setAllowedNetworkTypes(DownloadManager.Request.NETWORK_MOBILE | DownloadManager.Request.NETWORK_WIFI);
request.setAllowedOverRoaming(false);
//设置文件类型
MimeTypeMap mimeTypeMap = MimeTypeMap.getSingleton();
String mimeString = mimeTypeMap.getMimeTypeFromExtension(MimeTypeMap.getFileExtensionFromUrl(url));
request.setMimeType(mimeString);
```

```java
//在通知栏中显示
request.setNotificationVisibility(View.VISIBLE);
request.setVisibleInDownloadsUi(true);
File apkfile = new File(Environment.getExternalStorageDirectory() + File.separator + "download" + File.separator, name.concat(".apk"));
if (apkfile.exists())
    apkfile.delete();
//sdcard 目录中的 download 文件夹
request.setDestinationInExternalPublicDir("/download/", name.concat(".apk"));
request.setTitle(name);
long id = 0;
try
{
    id = downloadManager.enqueue(request);
} catch (Exception e)
{
    ApplicationUpdate mUpdate = new ApplicationUpdate(getBaseContext());
    HashMap<String, String> mHashMap = new HashMap<String, String>();
    mHashMap.put("name", name);
    mHashMap.put("url", url);
    mUpdate.showDownloadDialog(mHashMap);
}
SharedPreferences.Editor editor = prefs.edit();
editor.putString("name", name.concat(".apk"));
editor.putLong(BaseFragmentActivity.DL_ID, id);
editor.commit();
application.showToast(name + "开始下载。");
registerReceiver(receiver, new IntentFilter(DownloadManager.ACTION_DOWNLOAD_COMPLETE));
BaseFragmentActivity.receiver_state = 1;
```

若 Build.VERSION.SDK_INT 小于 Build.VERSION_CODES.HONEYCOMB_MR1，则调用系统下载器，代码如下：

```java
ApplicationUpdate mUpdate = new ApplicationUpdate(this);
HashMap<String, String> mHashMap = new HashMap<String, String>();
mHashMap.put("name", name);
mHashMap.put("url", url);
mUpdate.showNoticeDialog(name, content, mHashMap);
```

2）下载状态监听

用 Handler 的 handleMessage(Message msg)方法监听下载状态，代码如下：

```java
public Handler mHandler = new Handler()
{
```

```java
    public void handleMessage(Message msg)
    {
        switch (msg.what)
        {
        //正在下载
        case DOWNLOAD:
            //设置进度条位置
            mProgress.setProgress(progress);
            break;
        case DOWNLOAD_FINISH:
            SharedPreferences userInfo =mContext.getSharedPreferences
(MicroShopConfig.SHARE_TAG, Activity.MODE_PRIVATE);
            Editor editor =userInfo.edit();
            editor.remove("remember");
            editor.remove("user_psd");
            editor.commit();
                //安装文件
            installApk();
            break;
        default:
            break;
        }
    }
};
```

3) 下载对话框

利用 AlertDialog 可实现下载对话框,要处理好对话框单击事件和业务逻辑直接的关系。当 ProgressBar 出现时,需要同步更新下载进度。若采用系统自带的下载方式,则不用自己处理下载进度。

```java
public void showDownloadDialog(HashMap<String, String>mHashMap)
{
    this.mHashMap =mHashMap;
    //构造软件下载对话框
    AlertDialog.Builder builder =new Builder(mContext);
    builder.setTitle("下载更新");
    //给下载对话框增加进度条
    final LayoutInflater inflater =LayoutInflater.from(mContext);
    View v =inflater.inflate(R.layout.softupdate_progress, null);
    mProgress = (ProgressBar) v.findViewById(R.id.update_progress);
    builder.setView(v);
    //取消更新
    builder.setNegativeButton("取消更新", new OnClickListener()
    {
```

```
        @Override
        public void onClick(DialogInterface dialog, int which)
        {
            dialog.dismiss();
            //设置取消状态
            cancelUpdate = true;
        }
    });
    builder.setCancelable(false);
    mDownloadDialog = builder.create();
    mDownloadDialog.show();
    //下载文件
    downloadApk();
}
```

开始下载,代码如下:

```
//启动新线程下载文件
new downloadApkThread().start();
```

4) 将文件下载到 SD 卡

首先判断是否存在 SD 卡,以及 SD 卡的读写权限。若不进行判断,用户手机中 SD 卡不存在,或者 SD 卡损坏,还继续写入文件流,就会直接报错,影响用户体验。

```
//判断 SD 卡是否存在,并且是否具有读写权限
Environment.getExternalStorageState().equals(Environment.MEDIA_MOUNTED)
```

在 Manifest.xml 中,申请对 SD 卡读取的权限。

```
<uses-permission android:name="android.permission.READ_EXTERNAL_STORAGE" />
```

在 Manifest.xml 中,申请对 SD 卡的写入权限。

```
<uses-permission android:name="android.permission.WRITE_EXTERNAL_STORAGE" />
```

再在 SD 卡的指定位置创建文件,每次都进行判断和创建是为了防止用户手动清理过文件夹,导致程序崩溃。

```
//获得存储卡的路径
String sdpath = Environment.getExternalStorageDirectory() + File.separator +
"meDemo" + File.separator;
mSavePath = sdpath + "download";
```

从网络读取文件流,写入到创建的文件中。

```
URL url = new URL(mHashMap.get("url"));
//创建连接
HttpURLConnection conn = (HttpURLConnection) url.openConnection();
conn.connect();
```

```java
//获取文件大小
int length =conn.getContentLength();
//创建输入流
InputStream is =conn.getInputStream();
File file =new File(mSavePath);
//判断文件目录是否存在
if (!file.exists())
{
    file.mkdir();
}
File apkFile =new File(mSavePath, mHashMap.get("name").concat(".apk"));
FileOutputStream fos =new FileOutputStream(apkFile);
int count =0;
//缓存
byte buf[] =new byte[1024];
//写入到文件中
do
{
    int numread =is.read(buf);
    count +=numread;
    //计算进度条位置
    progress =(int) (((float) count / length) * 100);
    //更新进度
    mHandler.sendEmptyMessage(DOWNLOAD);
    if (numread <=0)
    {
        //下载完成
        mHandler.sendEmptyMessage(DOWNLOAD_FINISH);
        break;
    }
    //写入文件
    fos.write(buf, 0, numread);
} while (!cancelUpdate);//单击"取消"按钮就停止下载
```

最后，关闭文件流。

```java
fos.close();
is.close();
```

5）安装新版 APK

首先卸载当前栈所在的应用，再安装新版 APK。当前栈无法卸载当前栈的应用。flags 设置系统常量 Intent.FLAG_ACTIVITY_NEW_TASK。设置 intent.addFlags(Intent.FLAG_ACTIVITY_NEW_TASK)，将跳转后的 Activity 放到一个新的 Task（栈）中。

在安卓系统中，文件都有对应的文件 MIME 类型，不同的 MIME 类型通过 Intent 会调用不同的系统关联程序。此处，application/vnd. android. package-archive 会调用 APK 安装程序。

```java
public void installApk()
{
    File apkfile =new File(mSavePath, mHashMap.get("name").concat(".apk"));
    if (!apkfile.exists())
    {
        return;
    }
    //通过 Intent 安装 APK 文件
    Intent intent =new Intent();
    intent.addFlags(Intent.FLAG_ACTIVITY_NEW_TASK);
    intent.setAction(android.content.Intent.ACTION_VIEW);
    intent.setDataAndType(Uri.parse("file://" +apkfile.toString()),
"application/vnd.android.package-archive");
    mContext.startActivity(intent);
}
```

下面实现文件管理器。单击不同的文件，调用不同的程序。首先获取文件名的后缀名。

```java
//获取文件后缀名
private String getPrefix(File file)
{
    String fileName =file.getName();
    String prefix =fileName.substring(fileName.lastIndexOf(".") +1);
    return prefix;
}
```

然后建立一个 MIME 文件类型与文件后缀名的匹配表。

```java
private HashMap<String, String>MIMEMap =new HashMap<String, String>();
//建立一个 MIME 文件类型与文件后缀名的匹配表
private void initMIMEMap()
{
    //MIMEMap.put(后缀名，MIME 类型);
    MIMEMap.put(".3gp", "video/3gpp");
    MIMEMap.put(".apk", "application/vnd.android.package-archive");
    MIMEMap.put(".asf", "video/x-ms-asf");
    MIMEMap.put(".avi", "video/x-msvideo");
    MIMEMap.put(".bin", "application/octet-stream");
    MIMEMap.put(".bmp", "image/bmp");
    MIMEMap.put(".c", "text/plain");
    MIMEMap.put(".class", "application/octet-stream");
    MIMEMap.put(".conf", "text/plain");
```

```java
MIMEMap.put(".cpp", "text/plain");
MIMEMap.put(".doc", "application/msword");
MIMEMap.put(".exe", "application/octet-stream");
MIMEMap.put(".gif", "image/gif");
MIMEMap.put(".gtar", "application/x-gtar");
MIMEMap.put(".gz", "application/x-gzip");
MIMEMap.put(".h", "text/plain");
MIMEMap.put(".htm", "text/html");
MIMEMap.put(".html", "text/html");
MIMEMap.put(".jar", "application/java-archive");
MIMEMap.put(".java", "text/plain");
MIMEMap.put(".jpeg", "image/jpeg");
MIMEMap.put(".jpg", "image/jpeg");
MIMEMap.put(".js", "application/x-javascript");
MIMEMap.put(".log", "text/plain");
MIMEMap.put(".m3u", "audio/x-mpegurl");
MIMEMap.put(".m4a", "audio/mp4a-latm");
MIMEMap.put(".m4b", "audio/mp4a-latm");
MIMEMap.put(".m4p", "audio/mp4a-latm");
MIMEMap.put(".m4u", "video/vnd.mpegurl");
MIMEMap.put(".m4v", "video/x-m4v");
MIMEMap.put(".mov", "video/quicktime");
MIMEMap.put(".mp2", "audio/x-mpeg");
MIMEMap.put(".mp3", "audio/x-mpeg");
MIMEMap.put(".mp4", "video/mp4");
MIMEMap.put(".mpc", "application/vnd.mpohun.certificate");
MIMEMap.put(".mpe", "video/mpeg");
MIMEMap.put(".mpeg", "video/mpeg");
MIMEMap.put(".mpg", "video/mpeg");
MIMEMap.put(".mpg4", "video/mp4");
MIMEMap.put(".mpga", "audio/mpeg");
MIMEMap.put(".msg", "application/vnd.ms-outlook");
MIMEMap.put(".ogg", "audio/ogg");
MIMEMap.put(".pdf", "application/pdf");
MIMEMap.put(".png", "image/png");
MIMEMap.put(".pps", "application/vnd.ms-powerpoint");
MIMEMap.put(".ppt", "application/vnd.ms-powerpoint");
MIMEMap.put(".prop", "text/plain");
MIMEMap.put(".rar", "application/x-rar-compressed");
MIMEMap.put(".rc", "text/plain");
MIMEMap.put(".rmvb", "audio/x-pn-realaudio");
MIMEMap.put(".rtf", "application/rtf");
MIMEMap.put(".sh", "text/plain");
MIMEMap.put(".tar", "application/x-tar");
```

```
        MIMEMap.put(".tgz", "application/x-compressed");
        MIMEMap.put(".txt", "text/plain");
        MIMEMap.put(".wav", "audio/x-wav");
        MIMEMap.put(".wma", "audio/x-ms-wma");
        MIMEMap.put(".wmv", "audio/x-ms-wmv");
        MIMEMap.put(".wps", "application/vnd.ms-works");
        MIMEMap.put(".xml", "text/plain");
        MIMEMap.put(".z", "application/x-compress");
        MIMEMap.put(".zip", "application/zip");
        MIMEMap.put("", "*/*");
    };
```

Intent 动态传入 DataAndType,就可以方便地管理文件了。

```
intent.setDataAndType(Uri.parse("file://" + file.toString()), MIMEMap.get(getPrefix(file)));
```

12.6 知识点回顾

本章主要知识点如下:
(1) APK 应用打包与 Manifest.xml 的升级管理。
(2) 将 APK 发布到第三方应用市场。

第 13 章

与用户终端设备无关的 HTML 5

13.1 什么是 HTML 5

13.1.1 综述

HTML 5 是 HTML 主要的修订版本,现在仍处于发展阶段。其目标是取代 1999 年制定的 HTML 4.01 和 XHTML 1.0 标准,以期能在互联网应用迅速发展的时候,使网络标准符合当代需求。广义上,HTML 5 指的是包括 HTML、CSS 和 JavaScript 在内的一套技术组合。它希望减少浏览器对于需要插件的丰富的因特网应用服务(Rich Internet Application,RIA),如 Adobe Flash、Microsoft Silverlight 与 Oracle JavaFX 的需求,并且提供更多有效增强网络应用的标准集。

具体来说,HTML 5 添加了许多新的语法特征,其中包括＜video＞、＜audio＞和＜canvas＞元素,同时集成了 SVG 的内容。这些元素是为了更容易在网页中添加和处理多媒体和图片内容而添加的。其他新的元素包括＜section＞、＜article＞、＜header＞和＜nav＞,可丰富文档的数据内容。新的属性的添加也是为了同样的目的,同时也删除了某些属性和元素,＜a＞和＜menu＞也被修改、重新定义或标准化了。同时 APIs 和 DOM 已经成为 HTML 5 中的基础部分。HTML 5 还定义了处理非法文档的具体细节,使得所有浏览器和客户端程序能够统一处理语法错误。

13.1.2 发展历史

HTML 5 草案的前身是 Web Applications 1.0,2004 年由 WHATWG 提出,2007 年获 W3C 接纳,并成立了新的 HTML 工作团队。2008 年 1 月 22 日,第一份正式草案发布。WHATWG 表示该规范是目前仍在进行的工作,需不断完善。目前 Firefox、Google Chrome、Opera、Safari(版本 4 以上)、Internet Explorer(版本 9 以上)已支持 HTML 5 技术。

虽然网络开发人员已非常熟悉 HTML 5 了,但是它成为主流媒体的话题是在 2010 年 4 月。当时苹果公司的 CEO 乔布斯发表了一篇题为"对 Flash 的思考"的文章,指出"随着 HTML 5 的发展,观看视频或其他内容时,Adobe Flash 将不再是必需的。"这引发了开发人员间的争论。HTML 5 虽然提供了增强的功能,但开发人员必须考虑到不同浏览器对标准不同部分的支持程度,以及 HTML 5 和 Flash 功能的差异。

13.1.3 特性

1. 语义特性(Class：Semantic)

HTML 5 赋予网页更好的意义和结构。其更加丰富的标签将随着对 RDFa 的微数据与微格式等方面的支持,可构建对程序和用户都更有价值的数据驱动的 Web。

(1) 本地存储特性(Class：Offline & Storage)。HTML 5 开发的网页 APP 拥有更短的启动时间、更快的联网速度,这些全得益于 HTML 5 APP Cache 以及本地存储功能 Indexed DB(HTML 5 本地存储最重要的技术之一)和 API 说明文档。

(2) 设备兼容特性(Class：Device Access)。从具有 Geolocation 功能的 API 文档公开以来,HTML 5 为网页应用开发人员提供了更多功能上的优化选择。HTML 5 提供了前所未有的数据与应用接入开放接口,使外部应用可以与浏览器内部的数据直接相连,例如视频影音可直接与麦克风及摄像头相连。

(3) 连接特性(Class：Connectivity)。更有效的连接工作效率,使得基于页面的实时聊天、更快速的网页游戏体验、更优化的在线交流得到了实现。HTML 5 拥有更有效的服务器推送技术,Server-Sent Event 和 WebSockets 就是其中的两个特性。这两个特性能够实现服务器将数据"推送"到客户端的功能。

(4) 网页多媒体特性(Class：Multimedia)。支持网页端的 Audio、Video 等多媒体功能,与网站自带的 APPS、摄像头、影音功能相得益彰。

(5) 三维、图形及特效特性(Class：3D,Graphics & Effects)。基于 SVG、Canvas、WebGL 及 CSS3 的 3D 功能,用户会惊叹于在浏览器中所呈现的惊人的视觉效果。

(6) 性能与集成特性(Class：Performance & Integration)。没有用户会永远等待你的 Loading——HTML 5 通过 XMLHttpRequest2 等技术,解决了以前的跨域等问题,可使 Web 应用和网站在多样化的环境中更快速地工作。

(7) CSS3 特性(Class：CSS3)。在不牺牲性能和语义结构的前提下,CSS3 中提供了更多的风格和更强的效果。此外,较之以前的 Web 排版,Web 的开放字体格式(WOFF)也提供了更好的灵活性和可控制性。

2. 沿革

HTML 5 提供了一些新的元素和属性,例如<nav>(网站导航块)和<footer>。这种标签将有利于搜索引擎的索引整理,同时可更好地帮助小屏幕装置和视障人士使用。除此之外,它还为其他浏览要素提供了新的功能,如<audio>和<video>标记。

(1) 取消了一些过时的 HTML 4 标记。其中包括纯粹显示效果的标记,如<font>和<center>,它们已经被 CSS 取代。

HTML 5 吸取了 XHTML 2 的一些建议,包括用来改善文档结构的功能,比如,新的 HTML 标签 header、footer、dialog、aside、figure 等,将使内容创作者更容易创建文档,之前的开发者在实现这些功能时一般都使用 div。

（2）将内容和展示分离。b 和 i 标签依然保留，但其意义已经和之前有所不同。这些标签只是为了将一段文字标识出来，而不是为它们设置粗体或斜体式样。u、font、center、strike 标签则被完全去掉了。

（3）全新的表单输入对象。包括日期、URL、E-mail 地址，并增加了对非拉丁字符的支持。HTML 5 还引入了微数据，使用机器可以识别的标签标注内容的方法，使语义 Web 的处理更为简单。总之，这些与结构有关的改进使内容创建者可以创建更干净、更容易管理的网页，这样的网页对搜索引擎、读屏软件等更为友好。

（4）全新的、更合理的 Tag。多媒体对象将不再全部绑定在 Object 或 Embed Tag 中，而是视频有视频的 Tag，音频有音频的 Tag。

（5）本地数据库。内嵌一个本地的 SQL 数据库，以增强交互式搜索、缓存以及索引功能。同时，离线 Web 程序也将因此获益匪浅。

（6）Canvas 对象。用户可以脱离 Flash 和 Silverlight，直接在浏览器中显示图形或动画。

（7）浏览器中的真正程序。提供 API 实现浏览器内的编辑、拖放以及各种图形用户界面的能力。内容修饰 Tag 将被剔除，而使用 CSS。

（8）HTML 5 将取代 Flash 在移动设备中的地位。

（9）强化了 Web 页的表现性，追加了本地数据库。

13.1.4　未来趋势

HTML 5 的规范开发完成时，也许将成为市场应用的主流。

1．可能会消灭 Flash

许多业内人士认为，HTML 将会最终代替多媒体框架，如 Adobe 的 Flash，但是近期还不是时候，将现有应用 Flash 的网络开发完全转向 HTML 5 还需要一段时间。尽管 HTML 5 有许多优点，但是可能有某些应用适合于更灵活的框架。目前一些主流的大公司都逐步转向使用 HTML 5，但这个转变的过程不是一蹴而就的。

2．可能不够安全

HTML 5 所构建的网页和其他语言编写的网页一样容易泄露一些敏感数据。欧洲网络信息安全机构（European Network and Information Security Agency，ENISA）已经发出警告：HTML 5 可能不够安全。

3．承诺带来一个无缝的网络

HTML 5 会带来一个统一的网络，无论是笔记本、台式机，还是智能手机都可很方便地浏览基于 HTML 5 的网站。在设计网站的时候，开发者需要重新考虑用户体验、网站浏览、网站结构等因素，使得网站对任何硬件设备都通用。

4. 将会变成企业的 SaaS 平台

一些重量级的企业，如微软、Salesforce、SAP Sybase 正在开发 HTML 5 的开发工具。如果你正在构建企业应用，很可能不久的将来就要用到 HTML 5。因此当构建公司的 SaaS 战略迁移的时候，不要忘记 HTML 5。

5. 将会变得很移动

几乎所有人都热衷于开发独立的移动应用，但是 HTML 5 很可能会是独立移动应用的终结者。由于 HTML 5 将应用的功能直接加入其内核，因此可能会引导移动技术潮流重新回到浏览器时代。HTML 5 允许开发者在（移动）浏览器内开发应用，如果正在制定一项桌面或者移动应用的长期发展策略，就需要考虑到这一点。

6. 网络标准

HTML 5 本身是由 W3C 推荐的，它的开发是谷歌、苹果、诺基亚、中国移动等几百家公司一起酝酿的技术。其最大的好处在于它是公开的技术。换句话说，每一个公开的标准都可以根据 W3C 的资料库找寻根源。此外，W3C 通过的 HTML 5 标准也意味着每个浏览器或平台都会去实现。

7. 多设备跨平台

HTML 5 可以进行跨平台使用。比如 HTML 5 的游戏很容易移植到 UC 的开放平台、Opera 的游戏中心、Facebook 应用平台，甚至可以通过封装技术发放到 APP Store 或 Google Play 上。因此，HTML 5 的跨平台性非常强大，这也是大多数人对它有兴趣的主要原因。

8. 自适应网页设计

很早就有人设想，能不能"一次设计，普遍适用"，让同一张网页自动适应不同大小的屏幕，根据屏幕宽度，自动调整布局（layout）。

2010 年，Ethan Marcotte 提出了"自适应网页设计"这个概念，指可以自动识别屏幕宽度，并做出相应调整的网页设计。这就打破了传统的局面——网站为不同的设备提供不同的网页，比如专门提供一个手机版本，或者 iPhone、iPad 版本。这样做虽然保证了效果，但是比较麻烦，不仅同时要维护好几个版本，而且如果一个网站有多个入口（portal），就会大大增加架构设计的复杂度。

9. 即时更新

游戏客户端每次都要更新，很麻烦，然而更新 HTML 5 游戏就好像更新页面一样，是即时的更新。

综上所述，HTML 5 有以下优点：

（1）提高可用性和改进用户的友好体验。

(2) 有几个新的标签，这将有助于开发人员定义重要的内容。
(3) 可以给站点带来更多的多媒体元素(视频和音频)。
(4) 可以很好地替代 Flash 和 Silverlight。
(5) 涉及网站的抓取和索引时，对于 SEO 很友好。
(6) 可大量应用于移动应用程序和游戏开发。
(7) 可移植性好。

10．移动优先

在当今智能手机和平板电脑迅速发展的时代，移动应用层出不穷，移动优先已成趋势，不管开发什么产品，都以移动为主。

11．游戏开发者领衔"主演"

许多游戏开发商都被 Facebook 或者 Zynga 推动着发展，而未来的 Facebook 应用生态系统是基于 HTML 5 的。尽管在 HTML 5 平台开发游戏非常困难，但游戏开发商却都愿意这样做。通过 PhoneGap 及 appmobi 的 XDK 将 Web 应用游戏打包整合到原生应用中也是一种方式，Facebook 就是这样做的——基于 Web 应用及浏览器，但却将其打包整合进原生应用。

13.2 用 HTML 5 实现内容展示

13.2.1 WebView 组件

HTML 5 离不开浏览器，因此，需要比较全面地了解 Android 组件中的重要组件 WebView。

1．什么是 webkit

在 Android 手机中内置了一款高性能的 webkit 内核浏览器，webView 就是基于 webkit 的组件。

2．Android 与 JavaScript 之间的互相调用

WebView 默认禁用 JavaScript，在启用后，就可以在两者间建立接口进行调用。代码如下：

```
WebView myWebView = (WebView) findViewById(R.id.webview);
WebSettings webSettings =myWebView.getSettings();
webSettings.setJavaScriptEnabled(true);
```

webSetting 的功能非常强，可以开启很多设置，之后在本地存储、地理位置中都会用到。

3. 在 JavaScript 中调用 Android 函数的方法

在 Android 程序中建立接口，代码如下：

```java
final class InJavaScript {
    public void runOnAndroidJavaScript(final String str) {
        handler.post(new Runnable() {
            public void run() {
                TextView show = (TextView) findViewById(R.id.textview);
                show.setText(str);
            }
        });
    }
}

//把本类的一个实例添加到 js 的全局对象 window 中
//这样就可以使用 window.andMethod 来调用其方法了
webView.addJavascriptInterface(new InJavaScript(), "andMethod");
```

在 JavaScript 中调用，代码如下：

```javascript
function sendToAndroid(){
    var str = "调用 android 的方法";
    window.andMethod.runOnAndroidJavaScript(str);
}
```

4. 在 Android 中调用 JavaScript 的方法

JavaScript 中的方法：

```javascript
function getFromAndroid(str){
    document.getElementById("android").innerHTML=str;
}
```

在 Android 调用该方法：

```java
Button button = (Button) findViewById(R.id.button);
button.setOnClickListener(new OnClickListener() {
    public void onClick(View v) {
        //调用 JavaScript 中的方法
        webView.loadUrl("javascript:getFromAndroid('Cookie call the js function from Android')");
    }
}
```

要在 Android 中处理 JavaScript 的警告、对话框等，需要对 WebView 设置 WebChromeClient 对象。代码如下：

```java
//设置 WebChromeClient
webView.setWebChromeClient(new WebChromeClient(){
//处理 javascript 中的 alert
public boolean onJsAlert (WebView view, String url, String message, final JsResult result) {
//构建一个 Builder 来显示网页中的对话框
Builder builder =new Builder(MainActivity.this);
builder.setTitle("Alert");
builder.setMessage(message);
builder.setPositiveButton(android.R.string.ok,
    new AlertDialog.OnClickListener() {
        public void onClick(DialogInterface dialog, int which) {
            result.confirm();
        }
    });
    builder.setCancelable(false);
    builder.create();
    builder.show();
    return true;
};
//处理 javascript 中的 confirm
public boolean onJsConfirm (WebView view, String url, String message, final JsResult result) {
    Builder builder =new Builder(MainActivity.this);
    builder.setTitle("confirm");
    builder.setMessage(message);
    builder.setPositiveButton(android.R.string.ok,
        new AlertDialog.OnClickListener() {
        public void onClick(DialogInterface dialog, int which) {
            result.confirm();
        }
    });
    builder.setNegativeButton(android.R.string.cancel,
        new DialogInterface.OnClickListener() {
            public void onClick(DialogInterface dialog, int which) {
                result.cancel();
            }
        });
    builder.setCancelable(false);
    builder.create();
    builder.show();
    return true;
};
@Override
```

```java
//设置网页加载的进度条
public void onProgressChanged(WebView view, int newProgress) {
    MainActivity.this.getWindow().setFeatureInt(Window.FEATURE_PROGRESS, newProgress * 100);
        super.onProgressChanged(view, newProgress);
    }
    //设置应用程序的标题title
    public void onReceivedTitle(WebView view, String title) {
        MainActivity.this.setTitle(title);
        super.onReceivedTitle(view, title);
    }
});
```

5．Android 中的调试

通过 JavaScript 代码输出 log 信息，代码如下：

Js 代码：console.log("Hello World");
og 信息：Console: Hello World http://www.example.com/hello.html :82

在 WebChromeClient 中实现 onConsoleMesaage() 回调方法，让其在 LogCat 中打印信息。代码如下：

```java
WebView myWebView = (WebView) findViewById(R.id.webview);
myWebView.setWebChromeClient(new WebChromeClient() {
    public void onConsoleMessage(String message, int lineNumber, String sourceID) {
        Log.d("MyApplication", message + " -- From line "  + lineNumber + " of " + sourceID);
    }
});
```

或

```java
WebView myWebView = (WebView) findViewById(R.id.webview);
    myWebView.setWebChromeClient(new WebChromeClient() {
        public boolean onConsoleMessage(ConsoleMessage cm) {
            Log.d("MyApplication", cm.message() + " -- From line "          + cm.lineNumber() + " of "   + cm.sourceId() );
            return true;
        }
    });
```

13.2.2　HTML 5 本地存储

HTML 5 提供了两种客户端存储数据的新方法：localStorage 没有时间限制，

sessionStorage 针对一个 Session 的数据存储。

JavaScript 代码如下：

```html
<script type="text/javascript">
localStorage.lastname="Smith";
document.write(localStorage.lastname);
</script>
<script type="text/javascript">
sessionStorage.lastname="Smith";
document.write(sessionStorage.lastname);
</script>
```

WebStorage 的 API。JavaScript 代码如下：

```javascript
//清空 storage
localStorage.clear();
//设置一个键值
localStorage.setItem("yarin","yangfegnsheng");
//获取一个键值
localStorage.getItem("yarin");
//获取指定下标的键的名称(如同 Array)
localStorage.key(0);
//return "fresh" //删除一个键值
localStorage.removeItem("yarin");
//注意一定要在设置中开启
setDomStorageEnabled(true)
```

在 Android 中进行操作，Java 代码如下：

```java
//启用数据库
webSettings.setDatabaseEnabled(true);
String dir = this.getApplicationContext().getDir("database", Context.MODE_PRIVATE).getPath();
//设置数据库路径
webSettings.setDatabasePath(dir);
//若使用 localStorage,则必须打开
webSettings.setDomStorageEnabled(true);
//扩充数据库的容量(在 WebChromeClinet 中实现)
public void onExceededDatabaseQuota (String url, String databaseIdentifier, long currentQuota,    long estimatedSize, long totalUsedQuota, WebStorage.QuotaUpdater quotaUpdater) {
    quotaUpdater.updateQuota(estimatedSize * 2);
}
```

数据库的增、删、改、查，代码如下：

```javascript
function initDatabase() {
```

```javascript
try {
    if (!window.openDatabase) {
        alert('你的浏览器不支持数据库');
    } else {
        var shortName = 'MEDEMODB';
        var version = '1.0';
        var displayName = 'medemo db';
        var maxSize = 100000;
        YARINDB = openDatabase(shortName, version, displayName, maxSize);
        createTables();
        selectAll();
    }
} catch (e) {
    if (e == 2) {
        console.log("版本不匹配.");
    } else {
        console.log("未知错误" + e + ".");
    }
    return;
}
function createTables() {
    YARINDB.transaction(
        function (transaction) {
            transaction.executeSql('CREATE TABLE IF NOT EXISTS me_demo (id INTEGER NOT NULL PRIMARY KEY, name TEXT NOT NULL, desc TEXT NOT NULL);', [], nullDataHandler, errorHandler);
        }
    );
    insertData();
}
function insertData() {
    YARINDB.transaction(
        function (transaction) {
            //页面初始化完毕,加载数据
            var data = ['1', 'medemo one', 'I am medemo one'];
            transaction.executeSql("INSERT INTO me_demo (id, name, desc) VALUES (?, ?, ?)", [data[0], data[1], data[2]]);
        }
    );
}
function errorHandler(transaction, error) {
    if (error.code == 1) {
        //数据库中表已经存在
```

```javascript
    } else {
        console.log('Oops.  Error was ' +error.message + ' (Code ' +error.code +')');
    }
    return false;
}
function nullDataHandler() {
    console.log("SQL Query Succeeded");
}
function selectAll() {
    YARINDB.transaction(
        function (transaction) {
            transaction.executeSql("SELECT * FROM me_demo;", [], dataSelectHandler, errorHandler);
        }
    );
}
function dataSelectHandler(transaction, results) {
    //Handle the results
    for (var i =0; i <results.rows.length; i++) {
        var row =results.rows.item(i);
        var newFeature =new Object();
        newFeature.name =row['name'];
        newFeature.decs =row['desc'];

        document.getElementById("name").innerHTML ="name:" +newFeature.name;
        document.getElementById("desc").innerHTML ="desc:" +newFeature.decs;
    }
}
function updateData() {
    YARINDB.transaction(
        function (transaction) {
            var data =['medemo two', 'I am medemo two'];
            transaction.executeSql("UPDATE me_demo SET name=?, desc=? WHERE id =1", [data[0], data[1]]);
        }
    );
    selectAll();
}
function ddeleteTables() {
    YARINDB.transaction(
        function (transaction) {
            transaction.executeSql("DROP TABLE me_demo;", [], nullDataHandler, errorHandler);
```

```
            }
        );
        console.log("Table 'me_demo' has been dropped.");
    }
    function initLocalStorage() {
        if (window.localStorage) {
            textarea.addEventListener("keyup", function () {
                window.localStorage["value"] = this.value;
                window.localStorage["time"] = new Date().getTime();
            }, false);
        } else {
            alert("浏览器不支持本地存储");
        }
    }
    window.onload = function () {
        initDatabase();
        initLocalStorage();
    }
```

13.2.3　HTML 5 的地理位置服务

在 Manifest.xml 中添加权限。代码如下：

```
<uses-permission android:name="android.permission.ACCESS_FINE_LOCATION" />
<uses-permission android:name="android.permission.ACCESS_COARSE_LOCATION" />
```

在 HTML 5 中，通过 navigator.geolocation 对象获取地理位置信息。

常用的 navigator.geolocation 对象有以下三种方法。JavaScript 代码如下：

```
//获取当前地理位置
navigator.geolocation.getCurrentPosition(success_callback_function, error_callback_function, position_options)
//持续获取地理位置
navigator.geolocation.watchPosition(success_callback_function, error_callback_function, position_options)
//清除持续获取地理位置事件
navigator.geolocation.clearWatch(watch_position_id)
```

其中 success_callback_function 为成功之后处理的函数，error_callback_function 为失败之后返回的处理函数，参数 position_options 是配置项。

JavaScript 代码如下：

```
//定位
function get_location() {
    if (navigator.geolocation) {
        navigator.geolocation.getCurrentPosition(show_map, handle_error,
```

```javascript
        {enableHighAccuracy: false, maximumAge: 1000, timeout: 15000});
    } else {
        alert("你的浏览器不支持HTML 5地图功能");
    }
}
function show_map(position) {
    var latitude =position.coords.latitude;
    var longitude =position.coords.longitude;
    var city =position.coords.city;
    document.getElementById("Latitude").innerHTML ="latitude:" +latitude;
    document.getElementById("Longitude").innerHTML ="longitude:" +longitude;
    document.getElementById("City").innerHTML ="city:" +city;
}
function handle_error(err) {
    switch (err.code) {
        case 1:
            alert("未经授权");
            break;
        case 2:
            alert("网络连接失败,无法使用地图");
            break;
        case 3:
            alert("连接超时");
            break;
        default:
            alert("未知错误");
            break;
    }
}
```

附录 A

AndroidManifest.xml 中的权限

表 A.1 权限及说明

权 限	说 明
android.permission.INTERNET	允许程序打开网络 sockets
android.permission.KILL_BACKGROUND_PROCESSES	允许应用呼叫 kill_Background Processes 方法
android.permission.MANAGE_ACCOUNTS	允许程序管理账户列表(在账户管理者中)
android.permission.MASTER_CLEAR	目前还没有明确的解释
android.permission.MODIFY_AUDIO_SETTINGS	允许程序修改全局音频设置
android.permission.MODIFY_PHONE_STATE	允许修改电话状态,如电源、人机接口等
android.permission.MODIFY_FORMAT_FILESYSTEMS	允许格式化可移除的存储仓库的文件系统
android.permission.MOUNT_UNMOUNT_FILESYSTEMS	允许挂载和取消挂载文件系统
android.permission.PERSISTENT_ACTIVITY	允许程序设置其 Activity 显示
android.permission.PROCESS_OUTGOING_CALLS	允许程序监视、修改有关拨出电话
android.permission.READ_CALENDAR	允许程序读取用户日历数据
android.permission.READ_CONTACTS	允许程序读取用户联系人数据
android.permission.READ_FRAME_BUFFER	Android 读入帧缓冲数据程序
android.permission.READ_HISTORY_BOOKMARKS	允许应用读取(非写)用户浏览历史和书签
android.permission.READ_INPUT_STATE	允许程序返回当前按键状态
android.permission.READ_LOGS	允许程序读取底层系统日志文件
android.permission.READ_OWNER_DATA	允许程序读取所有者数据
android.permission.READ_PHONE_STATE	允许读取电话的状态
android.permission.READ_SMS	允许程序读取短信息(Allows an application to read SMS messages.)
android.permission.READ_SYNC_SETTINGS	允许程序读取同步设置

续表

权 限	说 明
android.permission.READ_SYNC_STATS	允许程序读取同步状态
android.permission.REBOOT	请求重新启动设备
android.permission.RECEIVE_BOOT_COMPLETED	允许程序接收 ACTION_BOOT_COMPLETED 广播在系统完成启动
android.permission.RECEIVE_MMS	允许程序监控接收 MMS 彩信、记录或处理
android.permission.RECEIVE_SMS	允许程序监控接收短信息、记录或处理
android.permission.RECEIVE_WAP_PUSH	允许程序监控接收 WAP PUSH 信息
android.permission.RECORD_AUDIO	允许程序录制音频
android.permission.RESTART_PACKAGES	允许程序重新启动其他程序(此值已不使用)
android.permission.SEND_SMS	允许程序发送 SMS 短信
android.permission.SET_ACTIVITY_WATCHER	Android 监控或控制 Activities
android.permission.SET_ALWAYS_FINISH	Android 控制是否处于后台时活动间接完成
android.permission.SET_DEBUG_APP	配置程序用于调试
android.permission.SET_ORIENTATION	允许底层访问设置屏幕方向和实际旋转角度
android.permission.SET_PREFERRED_APPLICATIONS	允许程序修改列表参数 PackageManager.addPackageToPreferred() 和 PackageManager.removePackageFromPreferred() 方法
android.permission.SET_PROCESS_LIMIT	允许设置最大的运行进程数量
android.permission.SET_TIME	允许应用设置系统时间
android.permission.SET_TIME_ZONE	允许程序设置系统时区时间
android.permission.SET_WALLPAPER	允许程序设置壁纸
android.permission.SET_WALLPAPER_HINTS	允许程序设置壁纸 hits
android.permission.SIGNAL_PERSISTENT_PROCESSES	允许程序请求发送信号到所有显示的进程中
android.permission.STATUS_BAR	允许程序打开、关闭或禁用状态栏及图标
android.permission.SUBSCRIBED_FEEDS_READ	允许程序访问订阅 RSS Feed 内容提供
android.permission.SUBSCRIBED_FEEDS_WRITE	系统暂时保留该设置
android.permission.SYSTEM_ALERT_WINDOW	允许程序打开窗口使用 TYPE_SYSTEM_ALERT,显示在其他所有程序的顶层
android.permission.UPDATE_DEVICE_STATS	允许应用更新设备资料信息

续表

权限	说明
android.permission.USE_CREDENTIALS	允许应用从管理器得到授权请求
android.permission.VIBRATE	允许访问振动设备
android.permission.WAKE_LOCK	允许使用 PowerManager 的 WakeLocks 保持进程在休眠时从屏幕消失
android.permission.WRITE_APN_SETTINGS	允许程序写入 API 设置
android.permission.WRITE_CALENDAR	允许程序写入但不读取用户日历数据
android.permission.WRITE_CONTACTS	允许程序写入但不读取用户联系人数据
android.permission.WRITE_EXTERNAL_STORAGE	允许应用写（非读）用户的外部存储器
android.permission.WRITE_GSERVICES	允许程序修改 Google 服务地图
android.permission.WRITE_HISTORY_BOOKMARKS	允许应用写（非读）用户的浏览器历史和书签
android.permission.WRITE_OWNER_DATA	允许程序写入但不读取所有者数据
android.permission.WRITE_SECURE_SETTINGS	允许应用写或读取当前系统设置
android.permission.WRITE_SETTINGS	允许程序读取或写入系统设置
android.permission.WRITE_SMS	允许程序写短信
android.permission.WRITE_SYNC_SETTINGS	允许程序写入同步设置

附录 B

Intent 和 Action 汇总

表 B.1　Uri、Action 及功能

Uri	Action	功　能
geo：latitude,longitude	Intent.ACTION_VIEW	打开地图应用程序并显示指定的经纬度
geo：0,0？q=street+address	Intent.ACTION_VIEW	打开地图应用程序并显示指定的地址
http://web_address	Intent.ACTION_VIEW	打开浏览器程序并显示指定的 URL
https://web_address	Intent.ACTION_VIEW	打开浏览器程序并显示指定的 URL
tel：phone_number	Intent.ACTION_CALL	打开电话应用程序并拨打指定的电话号码
tel：phone_number	Intent.ACTION_DIAL	打开电话应用程序并拨打指定的电话号码
voicemail：	Intent.ACTION_DIAL	打开电话应用程序并拨打指定语音邮箱的电话号码
plain_text	Intent.ACTION_WEB_SEARCH	打开浏览器程序并使用 Google 搜索

表 B.2　Intent、Action 及说明

Intent	Action	说　明
CALL_ACTION	android.intent.action.CALL	拨打电话，被呼叫的联系人在数据中指定
EMERGENCY_DIAL_ACTION	android.intent.action.EMERGENCY_DIAL	拨打紧急电话号码
DIAL_ACTION	android.intent.action.DIAL	拨打数据中指定的电话号码
ANSWER_ACTION	android.intent.action.ANSWER	处理拨入的电话
DELETE_ACTION	android.intent.action.DELETE	从容器中删除给定的数据
PICK_ACTION	android.intent.action.PICK	从数据中选择一个项目(item)，将被选中的项目返回

续表

Intent	Action	说 明
DEFAULT_ACTION	android.intent.action.VIEW	和 VIEW_ACTION 相同,是在数据上执行的标准动作
LOGIN_ACTION	android.intent.action.LOGIN	获取登录凭证
ALL_APPS_ACTION	android.intent.action.ALL_APPS	列举所有可用的应用
CLEAR_CREDENTIALS_ACTION	android.intent.action.CLEAR_CREDENTIALS	清除登录凭证(credential)
GET_CONTENT_ACTION	android.intent.action.GET_CONTENT	让用户选择数据并返回
EDIT_ACTION	android.intent.action.EDIT	为指定的数据显示可编辑界面
BUG_REPORT_ACTION	android.intent.action.BUG_REPORT	显示 activity 报告错误
SETTINGS_ACTION	android.intent.action.SETTINGS	显示系统设置,输入:无
WALLPAPER_SETTINGS_ACTION	android.intent.action.WALLPAPER_SETTINGS	显示选择墙纸的设置界面,输入:无
SENDTO_ACTION	android.intent.action.SENDTO	向 data 指定的接收者发送一个消息
VIEW_ACTION	android.intent.action.VIEW	向用户显示数据
PICK_ACTIVITY_ACTION	android.intent.action.PICK_ACTIVITY	选择一个 activity,返回被选择的 activity 的类(名)
RUN_ACTION	android.intent.action.RUN	运行数据(指定的应用),无论它(应用)是什么
INSERT_ACTION	android.intent.action.INSERT	在容器中插入一个空项(item)
ADD_SHORTCUT_ACTION	android.intent.action.ADD_SHORTCUT	在系统中添加一个快捷方式
WEB_SEARCH_ACTION	android.intent.action.WEB_SEARCH	执行 Web 搜索
SYNC_ACTION	android.intent.action.SYNC	执行数据同步
MAIN_ACTION	android.intent.action.MAI	作为主入口点启动,不需要数据
LABEL_EXTRA	android.intent.extra.LABEL	大写字母开头的字符标签,和 ADD_SHORTCUT_ACTION 一起使用
INTENT_EXTRA	android.intent.extra.INTENT	和 PICK_ACTIVITY_ACTION 一起使用时,说明用户选择的用来显示的 activity;和 ADD_SHORTCUT_ACTION 一起使用的时候,描述要添加的快捷方式

Intent	Action	说　　明
TEMPLATE_EXTRA	android.intent.extra.TEMPLATE	新记录的初始化模板
XMPP_DISCONNECTED_ACTION	android.intent.action.XMPP_DI	XMPP连接已经被断开
XMPP_CONNECTED_ACTION	android.intent.action.XMPP_CONNECTED	XMPP连接已经被建立
BATTERY_CHANGED_ACTION	android.intent.action.BATTERY_CHANGED	充电状态，或者电池的电量发生变化
TIME_TICK_ACTION	android.intent.action.TIME_TICK	当前时间已经变化（正常的时间流逝）
DATA_ACTIVITY_STATE_CHANGED_ACTION	android.intent.action.DATA_ACTIVITY	电话的数据活动（data activity）状态（即收发数据的状态）已经改变
DATA_CONNECTION_STATE_CHANGED_ACTION	android.intent.action.DATA_STATE	电话的数据连接状态已经改变
MESSAGE_WAITING_STATE_CHANGED_ACTION	android.intent.action.MWI	电话的消息等待（语音邮件）状态已经改变
SIGNAL_STRENGTH_CHANGED_ACTION	android.intent.action.SIG_STR	电话的信号强度已经改变
SERVICE_STATE_CHANGED_ACTION	android.intent.action.SERVICE_STATE	电话服务的状态已经改变
PHONE_STATE_CHANGED_ACTION	android.intent.action.PHONE_STATE	电话状态已经改变
PROVIDER_CHANGED_ACTION	android.intent.action.PROVIDER_CHANGED	更新将要（真正）被安装
FOTA_INSTALL_ACTION	android.server.checkin.FOTA_INSTALL	更新已经被确认，马上就要开始安装
FOTA_READY_ACTION	android.server.checkin.FOTA_READY	更新已经被下载，可以开始安装
FOTA_RESTART_ACTION	android.server.checkin.FOTA_RESTART	恢复已经停止的更新下载
MEDIA_SCANNER_STARTED_ACTION	android.intent.action.MEDIA_SCANNER_STARTED	开始扫描介质的一个目录
MEDIA_BAD_REMOVAL_ACTION	android.intent.action.MEDIA_BAD_REMOVAL	扩展介质（扩展卡）已经从SD卡插槽拔出，但是挂载点（mount point）还没解除（unmount）

续表

Intent	Action	说明
MEDIA_MOUNTED_ACTION	android.intent.action.MEDIA_MOUNTED	扩展介质被插入,而且已经被挂载
MEDIA_REMOVED_ACTION	android.intent.action.MEDIA_REMOVED	扩展介质被移除
MEDIA_UNMOUNTED_ACTION	android.intent.action.MEDIA_UNMOUNTED	扩展介质存在,但是还没有被挂载(mount)
MEDIA_SHARED_ACTION	android.intent.action.MEDIA_SHARED	扩展介质的挂载被解除(unmount),因为它已经作为USB大容量存储被共享
SCREEN_OFF_ACTION	android.intent.action.SCREEN_OFF	屏幕被关闭
SCREEN_ON_ACTION	android.intent.action.SCREEN_ON	屏幕已经被打开
FOTA_CANCEL_ACTION	android.server.checkin.FOTA_CANCEL	取消所有被挂起的(pending)更新下载
DATE_CHANGED_ACTION	android.intent.action.DATE_CHANGED	日期被改变
UMS_DISCONNECTED_ACTION	android.intent.action.UMS_DISCONNECTED	设备从USB大容量存储模式退出
CONFIGURATION_CHANGED_ACTION	android.intent.action.CONFIGURATION_CHANGED	设备的配置信息已经改变,参见Resources.Configuration
UMS_CONNECTED_ACTION	android.intent.action.UMS_CONNECTED	设备进入USB大容量存储模式
PACKAGE_REMOVED_ACTION	android.intent.action.PACKAGE_REMOVED	设备上删除了一个应用程序包
PACKAGE_ADDED_ACTION	android.intent.action.PACKAGE_ADDED	设备上新安装了一个应用程序包
NETWORK_TICKLE_RECEIVED_ACTION	android.intent.action.NETWORK_TICKLE_RECEIVED	设备收到了新的网络"tickle"通知
TIME_CHANGED_ACTION	android.intent.action.TIME_SET	时间已经改变(重新设置)
TIMEZONE_CHANGED_ACTION	android.intent.action.TIMEZONE_CHANGED	时区已经改变
FOTA_UPDATE_ACTION	android.server.checkin.FOTA_UPDATE	通过FOTA下载并安装操作系统更新

续表

Intent	Action	说明
STATISTICS_STATE_CHANGED_ACTION	android.intent.action.STATISTICS_STATE_CHANGED	统计信息服务的状态已经改变
WALLPAPER_CHANGED_ACTION	android.intent.action.WALLPAPER_CHANGED	系统的墙纸已经改变
PROVISIONING_CHECK_ACTION	android.intent.action.PROVISIONING_CHECK	要求 polling of provisioning service 下载最新的设置
STATISTICS_REPORT_ACTION	android.intent.action.STATISTICS_REPORT	要求 receivers 报告自己的统计信息
MEDIA_SCANNER_FINISHED_ACTION	android.intent.action.MEDIA_SCANNER_FINISHED	已经扫描完介质的一个目录
MEDIABUTTON_ACTION	android.intent.action.MEDIABUTTON	用户按下了"Media Button"
MEDIA_EJECT_ACTION	android.intent.action.MEDIA_EJECT	用户想要移除扩展介质（拔掉扩展卡）
CALL_FORWARDING_STATE_CHANGED_ACTION	android.intent.action.CFF	语音电话的呼叫转移状态已经改变
BOOT_COMPLETED_ACTION	android.intent.action.BOOT_COMPLETED	在系统启动后，这个动作被广播一次（只有一次）
LAUNCHER_CATEGORY	android.intent.category.LAUNCHER	Activity 应该被显示在顶级的 launcher 中
PREFERENCE_CATEGORY	android.intent.category.PREFERENCE	Activity 是一个设置面板（preference panel）
SAMPLE_CODE_CATEGORY	android.intent.category.SAMPLE_CODE	作为示例代码示例使用（不是正常用户体验的一部分）
FRAMEWORK_INSTRUMENTATION_TEST_CATEGORY	android.intent.category.FRAMEWORK_INSTRUMENTATION_TEST	用作框架测试的测试代码
SELECTED_ALTERNATIVE_CATEGORY	android.intent.category.SELECTED_ALTERNATIVE	对于被用户选中的数据，activity 是它的一个可选操作
BROWSABLE_CATEGORY	android.intent.category.BROWSABLE	能够被浏览器安全使用的 activities 必须支持这个类别
EMBED_CATEGORY	android.intent.category.EMBED	能够在上级（父）activity 中运行
DEFAULT_CATEGORY	android.intent.category.DEFAULT	如果 activity 是对数据执行缺省动作的一个选项，则需要设置这个类别

续表

Intent	Action	说　明
DEVELOPMENT_PREFERENCE_CATEGORY	android.intent.category.DEVELOPMENT_PREFERENCE	activity 是一个设置面板
ALTERNATIVE_CATEGORY	android.intent.category.ALTERNATIVE	activity 是用户正在浏览的数据的一个可选操作
UNIT_TEST_CATEGORY	android.intent.category.UNIT_TEST	应该被用作单元测试（通过 test harness 运行）
GADGET_CATEGORY	android.intent.category.GADGET	activity 可以被嵌入宿主 activity
WALLPAPER_CATEGORY	android.intent.category.WALLPAPER	activity 能够为设备设置墙纸
TAB_CATEGORY	android.intent.category.TAB	activity 应该在 TabActivity 中作为一个 tab 使用
HOME_CATEGORY	android.intent.category.HOME	主屏幕（activity），设备启动后显示的第一个 activity
TEST_CATEGORY	android.intent.category.TEST	作为测试目的使用，不是正常的用户体验的一部分

表 B.3　标志、值及描述

标　　志	值	描　　述
MULTIPLE_TASK_LAUNCH	8 0x00000008	和 NEW_TASK_LAUNCH 联合使用，禁止将已有的任务改变为前景任务（foreground）
FORWARD_RESULT_LAUNCH	16 0x00000010	如果这个标记被设置，而且被一个已经存在的 activity 用来启动新的 activity，则已有 activity 的回复目标（reply target）会被转移给新的 activity
NEW_TASK_LAUNCH	4 0x00000004	设置以后，activity 将成为历史堆栈中的第一个新任务（栈顶）
SINGLE_TOP_LAUNCH	2 0x00000002	设置以后，如果 activity 已经启动，而且位于历史堆栈的顶端，将不再启动（不重新启动）activity
NO_HISTORY_LAUNCH	1 0x00000001	设置以后，新的 activity 不会被保存在历史堆栈中

参 考 文 献

[1] Ed Burnette. Android 基础教程[M]. 张波,高朝勤,杨越,译. 北京:人民邮电出版社,2009.
[2] W. Frank Ableson Charlie Collins Robi Sen. Google Android 揭秘[M]. 张波,高朝勤,杨越,译. 北京:人民邮电出版社,2010.
[3] 靳岩,姚尚朗. Google Android 开发入门与实战[M]. 北京:人民邮电出版社,2009.
[4] 余志龙,陈昱勋,郑名杰,等. Google Android SDK 开发范例大全[M]. 北京:人民邮电出版社,2010.
[5] 梅尔. Android 高级编程[M]. 王鹏杰,霍建同,译. 北京:清华大学出版社,2010.